The Environment in World History

Throughout human history, it has been necessary for humans to work with and modify their environments in order to survive. However, in recent years, the environmental changes brought about by humans have come to rival those within nature itself, leading to an ecological crisis that has brought the environment to the top of the global political agenda.

Covering the past 500 years of global history, *The Environment in World History* examines the processes that have transformed the earth and have put growing pressure on natural resources. Chapters and case studies explore a wide range of issues, including:

- the hunting of wildlife and the loss of biodiversity in nearly every part of the globe;
- the clearing of the world's forests and the development of strategies to halt their decline;
- the degradation of soils, one of the most profound and unnoticed ways in which humans have altered the planet;
- the impact of urban-industrial growth and the deepening 'ecological footprints' of the world's cities;
- the pollution of air, land and water as the 'inevitable' trade-off for continued economic growth worldwide.

The Environment in World History also examines how European expansion overseas brought about a fundamental reorganisation of the world's ecology as contact between long-isolated continents was re-established. It offers a fresh environmental perspective on familiar world-history narratives of imperialism and colonialism, trade and commerce, and technological progress and the advance of civilisation, and will be invaluable reading for all students of world history and environmental studies.

Stephen Mosley is Senior Lecturer in History at Leeds Metropolitan University. His previous publications include *The Chimney of the World: A History of Smoke Pollution in Victorian and Edwardian Manchester* (2008).

Themes in World History

Series editor: Peter N. Stearns

The Themes in World History series offers focused treatment of a range of human experiences and institutions in the world-history context. The purpose is to provide serious, if brief, discussions of important topics as additions to textbook coverage and document collections. The treatments will allow students to probe particular facets of the human story in greater depth than textbook coverage allows and to gain a fuller sense of historians' analytical methods and debates in the process. Each topic is handled over time – allowing discussions of changes and continuities. Each topic is assessed in terms of a range of different societies and religions – allowing comparisons of relevant similarities and differences. Each book in the series helps readers deal with world history in action, evaluating global contexts as they work through some of the key components of human society and human life.

Gender in World History
Peter N. Stearns

Consumerism in World History
The Global Transformation of Desire
Peter N. Stearns

Warfare in World History
Michael S. Neiberg

Disease and Medicine in World History
Sheldon Watts

Western Civilization in World History
Peter N. Stearns

The Indian Ocean in World History
Milo Kearney

Asian Democracy in World History
Alan T. Wood

Revolutions in World History
Michael D. Richards

Migration in World History
Patrick Manning

Sports in World History
David G. McComb

The United States in World History
Edward J. Davies II

Food in World History
Jeffrey M. Pilcher

Childhood in World History
Peter N. Stearns

Religion in World History
John Super and Briane Turley

Poverty in World History
Steven M. Beaudoin

Premodern Travel in World History
Steven S. Gosch and Peter N. Stearns

Premodern Trade in World History
Richard L. Smith

Sexuality in World History
Peter N. Stearns

Globalization in World History
Peter N. Stearns

Jews and Judaism in World History
Howard N. Lupovitch

The Environment in World History
Stephen Mosley

The Environment in World History

Stephen Mosley

LONDON AND NEW YORK

First edition published 2010
by Routledge
2 Park Square, Milton Park, Abingdon, Oxon OX14 4RN

Simultaneously published in the USA and Canada by Routledge
270 Madison Ave, New York, NY 10016

Routledge is an imprint of the Taylor & Francis Group, an informa business

Typeset in Times New Roman by
Taylor & Francis Books
Printed and bound in Great Britain by
CPI Anthony Rowe, Chippenham, Wiltshire

British Library Cataloguing in Publication Data
A catalogue record for this book is available from the British Library

Library of Congress Cataloging in Publication Data
Mosley, Stephen.
The environment in world history / Stephen Mosley. – 1st ed.
p. cm. – (Themes in world history)
'Simultaneously published in the USA and Canada' – T.p. verso.
Includes bibliographical references and index.
1. Human ecology – History. 2. Nature – Effect of human beings on – History. 3. World history. 4. Human ecology – Case studies. 5. Nature – Effect of human beings on – Case studies. I. Title.
GF13.M67 2010
304.2 – dc22
2009031603

ISBN 10: 0-415-40955-1 (hbk)
ISBN 10: 0-415-40956-X (pbk)
ISBN 10: 0-203-85953-7 (ebk)

ISBN 13: 978-0-415-40955-1 (hbk)
ISBN 13: 978-0-415-40956-8 (pbk)
ISBN 13: 978-0-203-85953-7 (ebk)

Contents

Tables

Acknowledgements

Leeds Metropolitan University generously allowed me study leave to get the book underway. Thanks to Monika Büscher for reading the manuscript with a critical eye. I am also grateful to the editorial staff at Routledge for their help in bringing this project to completion. Finally, I want to thank Peter Stearns for asking me to write this book.

Chapter 1

Introduction

Environment and history

Throughout human history, people have played an active role in modifying their environments in order to survive. But for most of the past 4 million years, large-scale environmental transformations were produced by natural forces such as the drift of continents, volcanic eruptions and shifts in climatic conditions. These forces are still at work, but human-induced environmental changes have now begun to rival those of nature. According to the Millennium Ecosystem Assessment (2005), the first comprehensive global report on the health of the planet, humankind's ever-growing demands for natural resources are seriously damaging the ecosystem 'services' that support life. Of the twenty-four services it evaluated, such as fresh-water supplies, clean air, genetic resources and fisheries, no fewer than fifteen are being degraded or used unsustainably. Recent satellite images illustrate the dramatic impacts of human enterprise – farming, industry and urbanism – on forest, grassland, river and coastal ecosystems. These images reveal that croplands, dams, mines, roads, buildings and other developmental activities are reshaping the face of the earth at an unprecedented rate. While environmental change is inevitable, early *Homo sapiens* trod relatively lightly on the land. Today, our larger and more numerous 'ecological footprints' are clearly visible from space.

Since the late 1960s, space exploration has inadvertently helped to advance the cause of environmentalism. Iconic photographs taken by astronauts showing the earth as 'a sparkling blue-and-white jewel' suspended in the dark vastness of space heightened global environmental consciousness. The realisation that we all share one world began to take root. The 1960s were the seedtime for the modern environmental movement, particularly in North America and Western Europe. Inspired and energised by influential eco-writers of the day such as Rachel Carson, Barry Commoner, Paul Ehrlich and Garrett Hardin, the 'protest generation' forced the previously marginal issue of the environment into the political mainstream. On Earth Day 1970, for example, around 20 million people across America demonstrated against agricultural pesticides, industrial pollution, urban sprawl and other threats to the well-being of the planet. Three decades later, hundreds of millions of people in 183 countries participated in Earth Day 2000. Reacting to both growing public pressure and advancing scientific knowledge, during the same period a series of international conferences, from the Stockholm United Nations Conference on the Human Environment to the Rio Earth Summit, saw governments agree to work cooperatively towards the goal of environmental protection. But acrimonious wrangling between the developed nations of the North and the developing nations of the South over the management of natural resources and the control of pollution emissions has meant that the primary aim of 'sustainability' is proving difficult to achieve. Nonetheless, the interconnectedness and planetary scale of our current ecological crises – biodiversity loss, deteriorating ecosystem services, climate change – have moved the environment to the top of the global political agenda.

The rise of the environment as a political issue encouraged the emergence of an innovative new field of historical study: environmental history. Born out of the activism of the 1960s and 1970s, environmental history analyses the 'role and place of nature in human life'. Its primary goal is to reveal how human action and environmental change are intertwined. Nature, instead of being merely the backdrop against which the affairs of humans are played out, is recognised as playing an active role in historical processes. To grasp fully the complexities of human–environment relationships, historical research is generally carried out at four levels:

1. understanding the dynamics of natural ecosystems in time;
2. examining the interactions between environment, technology and the socio-economic realm;
3. inquiry into environmental policy and planning;
4. exploring changing cultural values and beliefs about nature.

This interpretive framework, based on Donald Worster's ambitious model for 'doing' environmental history, prompts scholars to make connections between the different levels of analysis. However, there are currently few works that link all parts effectively. Rather than constituting a rigid schema, research on all four levels is perhaps best viewed as a general programme for study. Pragmatically, the vast majority of practitioners have chosen to focus on only one or two levels, particularly environment, technology and socio-economics, and changing attitudes to nature – as will this short study.

In making both humans and non-humans actors in an evolving 'global ecodrama', some organisational recasting of analytical frameworks is essential – especially when it comes to accounting for the role of nature. One of the most challenging things about environmental history is its interdisciplinarity. Influenced by the holism of ecology, from the outset it has been an inclusive and collaborative endeavour. As well as historians, the field attracts scholars from a wide range of disciplines, from historical geography through to social anthropology and the natural sciences. Explorations of the ways in which climate, soils, forests, rivers and animals act as 'co-creators' of histories are blurring the traditional boundaries between the humanities and the sciences. To write nature into historical narratives, environmental historians must often work with both textual records and scientific data. Where documentary accounts are lacking – or unavailable – modern scientific techniques such as ice-core analysis, dendrochronology and palynology can unlock important information stored in 'nature's archives'. Ice cores, collected at both poles, provide a record of global carbon-dioxide levels and temperature fluctuations extending back hundreds of thousands of years. Dendrochronology, the study of tree rings, has been used to examine the relationship between agriculture and climatic cycles over the centuries. Palynology, the analysis of fossilised spores and pollen extracted from bogs, lake sediments and archaeological sites, can help to reconstruct histories of human settlement, forest clearance and crop introductions.

Furthermore, the natural sciences provide valuable heuristic metaphors. For example, the concept of metabolism – adopted from biology – has been used profitably by urban-environmental historians such as Joel Tarr in tracing the linkages between the city and the countryside. Drawing on the ideas of the ecologist Eugene Odum, Tarr has likened modern cities to 'parasitic' living organisms, dependent for their survival on inputs of clean air and water, fresh food, fossil fuels and construction materials, and the removal of harmful outputs of waste. The study of resource flows and waste emissions has begun to reveal the long-term impacts of urban living on the wider environment, especially after the Industrial Revolution, tracking the 'ecological footprints' of cities such as Paris, Pittsburgh, Chicago and Manchester deep into their own hinterlands and beyond. Undertaking research in environmental history, then, entails a willingness to press into service data and concepts from both the social and natural sciences. Although still undertheorised, the interdisciplinary ethos of eco-historical research is transcending the academic borders that separate 'nature from culture, science from history, matter from mind'.

Historical studies of socio-environmental change can be written on any scale, from the macro to the micro. But given that political boundaries and ecological boundaries rarely coincide, in environmental history the familiar organisational framework of the nation-state is not always an apt unit of analysis. Regional and local-level approaches, centred on different types of rural and urban 'ecosystems' – coastal, forests, grasslands, riverine, market towns and industrial cities – can often produce more coherent case studies of how societies and environments shape and reshape each other over time. Since the 1970s, work at the meso- and micro-levels has progressed rapidly and expanded worldwide, especially in North America, western Europe, South Asia, Africa and Australia. More recently, scholars have also begun to explore the history of human–nature interactions in Latin America, Russia,

China and Japan. In the space of three decades, a comprehensive literature spanning the globe has developed, encouraging a small but growing number of academics to internationalise their research and to write comparative macro-scale environmental history. As 'everything connects' in nature (climate change illustrates this point well), from an analytical point of view adopting a 'Big History' approach obviously makes good sense. Environmental transformations, while often differing widely from one region to another, are increasingly recognised to be interrelated: a host of regional-scale changes all over the world cumulatively placing unsustainable pressure on the biosphere (the ecosystem of the entire planet). World-environmental historians, by drawing attention to the international transfer of plants, animals, microbes, agricultural practices and industrial technologies, as well as global patterns of trade, migration and settlement, have begun to trace the common underlying causes of ecological problems at both the regional and biospheric levels.

Some of the most important and interesting work now being done in world history is concerned with understanding and contextualising environmental change. This book aims to make the results of recent research in this area accessible to students and their instructors. At this point, however, one or two caveats about the project are in order. In a short, synthetic work of this type, it is not possible to cover the whole range of human–environment relationships over time and space. Rather, the book will survey several of the main themes in world-environmental history – deforestation, species loss, soil erosion and the pollution of air, land and water – using case studies drawn from Africa, England, India and North America to illustrate the bigger picture. The need to be concise also means that there is little discussion here of natural disasters unprovoked by human activity, such as earthquakes, volcanic eruptions, or El Niño events. What follows then charts the history of human interactions with nature from circa 1500 to the present (with the occasional look further backward) – a period when patterns of environmental change began to diverge significantly in scale and intensity from those of the past.

Periodisation and patterns of environmental change

Studying the relationship between humankind and the environment can also mean having to rethink systems of periodisation, which in environmental history are often defined by natural processes rather than conventional political markers. Many important works deal in 'deep time', beginning, for example, with the breakup of the supercontinent Pangaea – from which the present continents were eventually formed – some 180 million years ago. Timescales routinely encompass millennia and even geological eras, challenging historians to engage with chronologies that reduce the human lifespan to the mere blink of an eye. In addition, exploring history in deep time unquestionably provides us with a more humble view of the human role in historical processes. Writing over fifty years ago, Fernand Braudel stressed the slow rate of change in human–nature relations over *la longue durée* in his magisterial history of the Mediterranean, a history shaped by the ever-recurring cycles of the seasons, the deep-seated rhythms of rural life and the interdependence of people and place. To a considerable extent he saw nature as stable and unchanging, setting the limits to growth for human societies through its influence on land use and economic organisation. But more recently, in the light of growing threats to the sustainability of the ecosystem services we depend on, and new scientific insights that reveal nature to be unstable and chaotic, environmental historians have tended to privilege change over continuity in their work.

The choice of a relatively short temporal framework for this volume, taking up the story roughly where Braudel left off in the sixteenth century, focuses our attention on a particularly turbulent period in the changing human–nature relationship. When viewed over the past five centuries, rather than a very *longue durée* perspective, it is the speed with which humankind has altered the 'faces and flows' of the global environment that is remarkable. Starting with Columbus's 'discovery' of the New World in 1492, by the late eighteenth century European explorers had re-established contact between the long-isolated ecosystems and peoples of the Americas, Afro-Eurasia, Australasia and the Pacific islands. European expansion overseas brought about a fundamental reorganisation of the world's ecology, an Environmental Revolution, which affected the everyday lives of most of its inhabitants. Transoceanic exchanges – the increasing traffic in plants, animals, microbes and people around the world – were to alter the face of the earth.

The ecological impacts of early globalisation are difficult to overestimate. Restoring contact between the previously isolated Old and New Worlds exposed millions of people to unfamiliar disease-causing microbes to which they had no genetic resistance or acquired immunity. Deadly 'virgin soil' epidemics of smallpox, measles, influenza and other Eurasian diseases reduced Amerindian, Australian Aborigine, New Zealand Maori and Pacific island populations to a fraction of their former numbers. Although figures are hotly disputed, academics generally agree that indigenous New World populations suffered exceptionally high levels of mortality as a consequence of 'microbial unification': 90 per cent or more in some cases. One commentator has conservatively estimated that infectious diseases unwittingly spread by European exploration, trade and colonisation claimed around 56 million lives worldwide. However, with the possible exception of syphilis, the confluence of disease pools that followed the 'Columbian exchange' and the later 'Cook exchange' introduced no agents of disease that significantly raised European mortality rates.

Unplanned 'biological warfare' cleared the way for European settlers to reshape New World landscapes and to make them more like the Old. They planted wheat, barley, rice and other customary food crops in depopulated colonies and established large herds of domesticated European animals such as cattle, sheep, goats and horses. Lacking natural predators or aggressive competitors, most flourished in what Alfred Crosby called the 'lands of demographic takeover': Argentina, Australia, southern Brazil, Uruguay, New Zealand and much of North America. Wholesale introductions of Old World species crowded out native flora and fauna, resulting in both the homogenisation and simplification of ecosystems. By 1900, the temperate zone regions of the Americas and Australasia had been transformed by settlers on a grand scale into 'neo-Europes': biological replicas of their homelands. Just as importantly, from the sixteenth century onwards, highly productive New World food crops such as maize (corn), potatoes (white and sweet) and cassava (manioc) also crossed the oceans and spread vigorously throughout Afro-Eurasia, providing the nutritional foundations for population growth (especially in Europe and China).

Humans have moved species around for as long as they have migrated. But in the early modern period, seaborne European empires – Iberian, Dutch, French and English – both accelerated the pace and extended the range of biotic exchanges. Oceans became 'intercontinental highways' for the world's flora, fauna and microbes, rather than barriers to their movement, making possible rapid environmental change on a genuinely global scale. The Age of Sail brought with it a terrific expansion of the human capacity to alter the visible face of virtually every terrestrial ecosystem from, for instance, forest cover into animal pasture, or grassland environments into wheat fields. Land-use change is still ongoing and, to date, somewhere between 35 to 50 per cent of the earth's land surface has been radically transformed by human activity. Yet the single most important development within our chosen timeframe was the harnessing of fossil fuels, the environmental impacts of which – while often less obvious – were more profound, causing potentially catastrophic changes in the bio-geo-chemical flows that sustain the biosphere.

The Industrial Revolution, stimulated in part by European success in turning 'empty' landscapes around the world into productive agro-ecosystems, marked the transition from traditional to modern societies and economies. In the late eighteenth century, humankind began to break free from the constraints of what has been called the 'biological *ancien regime*'. Traditional agrarian societies relied mainly on renewable energy and resources – wood, water, wind, human and animal muscle – to earn a very modest living from the land for all but a privileged few. Forests kept households supplied with firewood and construction materials, while charcoal was widely utilised for smelting metals to produce tools, utensils and other handcrafted goods. Textile industries such as linen, cotton and wool depended on the products of fields and pasture. Not least of all, traditional farming required high labour inputs, fertile soils and clement weather to produce the bountiful harvests that were the key to human well-being. Life's necessities – food, fuel, clothing and shelter – all came from the land – land worked primarily by human muscle-power, with some limited assistance from windmills, waterwheels and draught animals. As people were unable to capture and convert into useful forms more than a fraction of the energy available on earth, within the biological *ancien régime* there were limits on both the size of populations and the productivity of the economy. The switch to fossil fuels during the industrial era changed all that.

The invention of an efficient steam engine in the late eighteenth century, the electric motor and the internal combustion engine in the nineteenth century, and the gas turbine in the early twentieth century, lessened people's dependence on wood, water, wind and muscle power and made possible the exponential growth of economies and populations. These technological innovations, prime movers of the Industrial Revolution, promoted and required the harvesting of a 'new' subterranean fossil-fuel 'acreage': coal-fields, oilfields and gasfields. Hitherto unimaginable stores of energy, accumulated over hundreds of millions of years, suddenly became available for human use. Agricultural societies had to meet their needs from within a finite acreage of woodland and farmland. Even with good farming techniques, recurring energy and resource shortages impeded their development. For example, clearing forests for agricultural use reduced the amount of wood available for fuel, construction and making charcoal. As long as economies were mainly based on muscle-driven agriculture and biomass energies their productive capacity remained relatively low, while the share of the population who were poor and rural remained high. Access to inexpensive and seemingly inexhaustible fossil fuels, albeit unevenly distributed around the world, broke the shackles of the biological *ancien régime* and restructured ways of living. Originating in Britain, industrialisation spread throughout Europe, North America, Japan and then the wider world. Drawing on energy stocks deposited in prehistory, the Industrial Revolution allowed modern societies to replace animate power with mechanical power and accelerated the pace of resource extraction, production and consumption.

The most visible sign of escalating economic activity was the rise of great industrial cities and agglomerations such as Manchester, Pittsburgh and Germany's Ruhrgebiet, based on the large-scale production of cotton textiles, iron and steel. There were few major cities within pre-industrial societies, their size and number held in check by the low productivity of their hinterlands. In 1800, no more than 7 per cent of the world's population lived in urban environments. Today, the proportion of the world's population living in cities has passed 50 per cent. In the booming cities of the Industrial Revolution, factories and furnaces vastly improved productivity over artisanal levels, supplied a growing range of affordable, high-quality goods, and brought a measure of affluence for most urban workers. There was an explosion of consumerism after 1850, symbolised by the department store, radiating out from the affluent cities of north-western Europe and the USA. Industrialisation also imposed a new rhythm on social life, with machines and clock time, rather than the natural cycles of days and seasons, dictating the quickening pace. In agriculture, labour-saving machinery released rural workers from the land, and artificial fertilisers (derived from fossil fuels) brought higher and more reliable crop yields, which underpinned ongoing urban-industrial growth. Over the past two centuries then, the control of fossil energies has enabled *Homo sapiens* to become a predominantly urban species – but at a high cost to the environment.

Industrialisation allowed humankind to use fossil fuels and other resources at prodigious rates. Table 1.1 below shows the dates by which 25 per cent, then 50 per cent, then 75 per cent of various human-induced environmental impacts were reached (calculations assume a baseline of zero for 10,000 BCE, and 100 per cent change as of 1985).

In recent centuries, industrial society has altered not only the faces of the earth, but also the flows of key elements – particularly phosphorous, nitrogen, sulphur and carbon – through the biosphere, with serious ecological consequences. Forest clearances have transformed land cover over vast areas and have accelerated species extinctions as suitable habitat shrank. Increasing freshwater withdrawals from rivers and lakes for irrigation, industrial and household use have disrupted the hydrological cycle. At times, the flows of major rivers, such as the Nile in Africa, the Colorado in North America and the Yellow River in China, are so reduced that they no longer reach the ocean. The intensifying use of phosphorous and nitrogen in artificial fertilisers, which in the latter case far outweighs natural releases, has caused widespread water pollution and eutrophication (where excess nutrients in agricultural runoff stimulate algal blooms that in turn reduce oxygen in the water, killing other forms of aquatic life). Rising sulphur emissions from smoke-stack industries and power generation have polluted the air and contributed

Table 1.1 Quartiles of human-induced environmental change from 10,000 BCE to 1985

Form of Transformation	*Dates of Quartiles*		
	25%	*50%*	*75%*
Deforestation	1700	1850	1915
Terrestrial vertebrate extinctions	1790	1880	1910
Water withdrawals	1925	1955	1975
Phosphorus releases	1955	1975	1980
Nitrogen releases	1970	1975	1980
Sulfur releases	1940	1960	1970
Carbon releases	1815	1920	1960

Source: adapted from Robert W. Kates, B. L. Turner and William C. Clark, 'The Great Transformation' in B. L. Turner, William C. Clark, Robert W. Kates, John F. Richards, Jessica T. Mathews and William B. Meyer (eds), *The Earth as Transformed by Human Action* (Cambridge: Cambridge University Press, 1990).

greatly to the emergence of an acid-rain problem in Europe and North America by the mid-1980s. Acid rain was a fast-growing problem in East Asia too by the mid-1990s (nowhere faster than in industrialising China).

The increase in the amount of carbon dioxide in the atmosphere, due primarily to burning fossil fuels, is potentially the most damaging environmental change. Global statistics reveal the exponential growth in fossil-fuel use since the advent of the Industrial Revolution. Coal production increased 500-fold between 1810 and 1990, from 10 million to 5 billion tonnes. Oil extraction rose around 300-fold in the century after 1890, from less than 10 million to more than 3 billion tonnes. Natural-gas production increased about 1,000-fold during the same period, from less than 2 billion to almost 2 trillion cubic metres. By living off the accumulated energy capital of the past instead of 'current income' (renewable energies), humankind released huge quantities of carbon dioxide that had been securely locked up underground, profoundly altering the global carbon cycle. Ice-core data from Greenland and Antarctica show conclusively that atmospheric concentrations of the 'greenhouse gas' carbon dioxide now far exceed pre-industrial values, up from around 280 to 379 parts per million in 2005 (an increase of over 35 per cent). Global warming, estimated at between 1.8 to 6.4 degrees Celsius by the end of the twenty-first century, will result in a whole range of 'natural disasters' that include desertification, droughts, species extinctions, sea-level rises and a destructive change in weather patterns.

As well as changing global elemental flows at a speed and on a scale that was impossible before industrialisation, humans have introduced numerous substances not previously found in nature: most notably oil-derived plastics, nuclear wastes and chlorofluorocarbons (the chemicals responsible for the Antarctic 'ozone hole'). In a relatively short space of time, humans have become a major influence on the biosphere and how it functions. Although emphases inevitably differ, recent studies have highlighted the following as the main human drivers of environmental change:

- **Population growth.** Between 1500 and 2000, the world's population grew from around 500 million to over 6 billion people (with most of this spectacular increase coming in the twentieth century). As numbers rose, so did demands for life's necessities – food, water, fuel, clothing and shelter – generally at the expense of supporting ecosystems. Population redistribution from the countryside to the city saw more and more people lose intimate contact with nature.
- **Technological advance.** After 1492, advances in transportation and communications played a crucial role in the transformation of the earth, facilitating the global movement of raw materials, trade goods, information, capital, people and other organisms. But it was the Industrial Revolution that decisively shifted the balance of power in the nature–human relationship. In the past two centuries, technological innovation

and cheap energy have allowed humans to shape their environments more than they have shaped us. Technology was, and still remains, the 'inexhaustible resource'; extracting wealth from the earth and circumventing energy and resource shortages. Modern industrial technologies help to support unprecedented numbers of people and unprecedented levels of material consumption – but the pressures they place on the planet's ecosystem 'services' are now becoming unsustainable.

- **Economic expansion.** Between 1500 and the early 1990s, the world's annual gross domestic product grew 120-fold, from around $240 billion to $28 trillion (with most growth coming after 1820). Europe's ability to exploit the natural resources of its colonies, reconfiguring ecologies and economies worldwide to meet the demands of industrial cities, lay behind much of this enormous increase in wealth and output. But the headlong pursuit of prosperity was undertaken with scant regard for the long-term conservation of natural resources (often finite), and disproportionately heavy environmental burdens (deforestation, soil erosion, loss of biodiversity) were borne by the poor countries that exported raw materials to the rich industrialised nations. Nor were the fruits of industrial capitalism shared equally. In 2001, just over 1 billion people in the developing world survived on less than $1 per day.
- **Cultural attitudes to nature.** From the sixteenth century, the world of nature was reconceptualised as machine-like to meet the needs of an emergent capitalism. In place of the notion that the earth was a single living organism – a cosmology shared by many of the world's peoples – capable of retribution against those who carried out destructive acts, this new scientific worldview saw nature as dead matter for human utilisation. Cartesian science and the rise of capitalism were highly compatible with Europe's leading religion, Christianity, which emphasised the theme of human domination of nature. While some writers have argued that Chinese, Indian, African and indigenous religions were more respecting of nature, non-Western cultures also recast and despoiled their environments in order to grow their economies and improve living standards, as did the modern socialist economies of the Soviet Union and Eastern Europe. Despite their differences, cultures everywhere refashioned nature to make it better serve humanity.

Over the past five centuries, the impacts of human activities on the 'faces and flows' of the global environment have escalated. Because of our extraordinary capacity to modify the natural world to meet our needs, the Nobel Laureate Paul Crutzen has suggested assigning the term 'Anthropocene' (the human epoch) to describe the present period of the earth's history. We will explore humankind's growing influence on the environment more fully in the chapters that follow.

Chapter 2 examines advancing hunting frontiers around the world, different cultural practices and attitudes to hunting, the economic advantages derived from hunting, efforts to conserve declining stocks and some of the environmental changes brought about by species loss. Chapter 3 considers the complex set of socio-cultural, technological and economic forces that have caused the world's forests to decline. It will also chart the growth and development of strategies to manage and conserve the earth's remaining temperate and tropical forests. Chapter 4 examines changing human relationships with the earth's soils, which are crucial for food production. Soil conservation has been – and still is – a major challenge for most societies, as it is a natural resource that is not renewable in the short term. Chapter 5 explores the impacts of urban-industrial growth on the global environment. As concentrated centres of population, production and consumption, cities have always placed heavy demands on natural resources. A focus on resource flows and waste emissions will reveal the long-term environmental effects of urban living, especially after the Industrial Revolution, tracking the 'ecological footprints' of major cities such as Manchester – the first industrial city – far beyond their own hinterlands. The concluding chapter highlights how local and regional environmental issues can connect to become genuinely global in scale.

Further reading

David Arnold, *The Problem of Nature: Environment, Culture and European Expansion* (Oxford: Blackwell, 1996).

Fernand Braudel, *The Mediterranean and the Mediterranean World in the Age of Philip II* (London: Collins, 1972).

Alfred Crosby, *Ecological Imperialism: The Biological Expansion of Europe, 900–1900* (Cambridge: Cambridge University Press, 2004).

Paul J. Crutzen, 'Geology of Mankind: The Anthropocene', *Nature* (vol. 415, January 2002, p. 23).

Roger S. Gottlieb (ed.), *This Sacred Earth: Religion, Nature, Environment* (New York: Routledge, 1996).

Ramachandra Guha, *Environmentalism: A Global History* (Harlow: Longman, 2000).

Rashid Hassan, Robert Scholes and Neville Ash (eds), *Millennium Ecosystem Assessment, Volume 1: Ecosystems and Human Well-Being – Current State and Trends* (Washington, DC: Island Press, 2005).

J. Donald Hughes (ed.), *The Face of the Earth: Environment and World History* (Armonk, NY: M. E. Sharpe, 2000).

Shephard Krech, John McNeill and Carolyn Merchant (eds), *Encyclopedia of World Environmental History* (New York: Routledge, 2004).

Tony McMichael, *Human Frontiers, Environments and Disease: Past Patterns, Uncertain Futures* (Cambridge: Cambridge University Press, 2001).

John McNeill, *Something New under the Sun: An Environmental History of the Twentieth Century* (London: Allen Lane, 2000).

John McNeill and William McNeill, *The Human Web: A Bird's-Eye View of World History* (New York: W. W. Norton, 2003).

Carolyn Merchant, *Radical Ecology: The Search for a Livable World* (New York: Routledge, 2005).

Robert B. Marks, *The Origins of the Modern World: A Global and Ecological Narrative* (Lanham, Md.: Rowman & Littlefield, 2002).

Timo Myllyntaus and Mikko Saikku (eds), *Encountering the Past in Nature: Essays in Environmental History* (Athens, Ga.: Ohio University Press, 1999).

Dieter Schott, Bill Luckin and Geneviève Massard-Guilbaud (eds), *Resources of the City: Contributions to an Environmental History of Modern Europe* (Aldershot: Ashgate, 2005).

Vaclav Smil, *Energy in World History (*Boulder, Col.: Westview, 1994).

Joel A. Tarr, 'The Metabolism of the Industrial City', *Journal of Urban History* (vol. 28, July 2002, pp. 511–45).

B. L. Turner, William C. Clark, Robert W. Kates, John F. Richards, Jessica T. Mathews and William B. Meyer (eds), *The Earth as Transformed by Human Action* (Cambridge: Cambridge University Press, 1990).

Ashbindu Singh, Thomas R. Loveland, Mark Ernste, Kimberley A. Giese, Rebecca L. Johnson, Jane S. Smith, John Hutchinson, Eugene Fosnight, H. Gyde Lund, Tejaswi Giri, Jane Barr, Eugene Apindi Ochieng and Audrey Ringler (eds), *One Planet, Many People: Atlas of Our Changing Environment* (United Nations Publications, 2005).

Donald Worster (ed.), *The Ends of the Earth: Perspectives on Modern Environmental History* (Cambridge: Cambridge University Press, 1988).

Chapter 2

The world hunt

Twenty-first century landscapes and ecosystems are undeniably impoverished in comparison with those of the early modern centuries, especially in terms of the declining range and number of large animal and marine species. Five hundred years ago, the earth's land surface and oceans teemed with an abundance and diversity of wildlife almost inconceivable today. Maps still show place names that capture something of this former richness, for instance, Rhenoster (rhinoceros) Kop in South Africa and Cape Cod in North America. But from the late fifteenth century, European colonial expansion and early forms of economic globalisation encouraged the rapid growth of what historian John Richards has called the 'world hunt'. Unrestrained commercial predation for furs, feathers, ivory and flesh killed off enormous numbers of the planet's wildlife, greatly diminishing the complexity of its ecosystems. In particular, populations of large mammals such as rhinos, tigers, elephants, bison and whales – whose carcasses possessed significant market value – declined dramatically. Since 1600, at least 485 animal species have become extinct (an extinction rate far higher than the 'normal' background average), many as a direct result of overhunting.

Until recently, the importance of hunting to the European colonial enterprise had not been fully recognised by historians. However, the work of Richards and others has shown that advancing colonial frontiers around the world were also hunting frontiers, with the rising demand for animal resources one of the primary drivers of expansion. The world hunt, as we shall see, comprised various types and overlapping phases of predation by humans, most notably organised commercial hunting for international markets, settler hunting for subsistence and vermin reduction and the killing of animals for sport and recreation. Over the past 500 years, systematic hunting on an unprecedented scale and of previously unseen intensity seriously depleted populations of what conservationists now call 'charismatic megafauna'. Most large wild animals disappeared from increasingly homogenised landscapes, replaced by a smaller roster of domesticated and synanthropic species – horses, cows, sheep, pigs, rats, house sparrows and so on – that thrive in human-dominated environments. While the earth still teems with life, it is much less exotic and diverse.

Humans have been hunter-gatherers for most of their evolutionary history, and some small isolated groups have persisted in remote areas up to the present day. The apparent success of hunter-gatherers in using natural resources wisely over millennia and allowing underlying environmental support systems to remain in good working order has generated considerable scholarly debate as to whether such groups were the 'first ecologists' (and hence possible role models for a more sustainable future). The Inuit peoples of Arctic North America, for example, developed a sustainable hunting culture that depended on marine mammals – whales, walruses and seals – as a major source of subsistence. Southern African societies, such as the San 'Bushmen', hunted for food and trade without noticeably disrupting the population dynamics of the vast numbers of wild animals dispersed throughout the region. There is, however, compelling archaeological evidence to suggest that some hunter-gatherers have been responsible for numerous cases of 'overkill' and species loss since the Pleistocene colonisation of the earth by humans. Whether or not hunter-gatherer societies, with their intimate knowledge of local ecosystems, were 'prudent predators' is still open to question. But their basic technologies, highly mobile lifestyles, lack of demand for material possessions and low population levels meant that their ecological impact was relatively light. In stark contrast, from the beginning of the early modern period, market-oriented hunting greatly accelerated the loss of biodiversity in nearly every part of the globe. It also enmeshed indigenous peoples in unequal trade relations with Europeans that both transformed their cultures and undermined their autonomy.

Commercial hunting

Market hunting for wildlife products went hand in hand with European exploration and expansion in the early modern period. Thanks to innovations in maritime transport, advances in mapping and navigational skills and the demands of a growing global economy, from the late fifteenth century organised commercial hunting was extended into every world region and ocean. Wealthy Europeans and Asians took pleasure in purchasing decorative objects fashioned from ivory, in wearing furs and feathers and in eating exotic cuts of meat and fish. Once a monetary value had been placed on these and other wildlife products – a new departure from pre-colonial trading systems that turned nature into a commodity – human hunters quickly depleted profitable 'target species' in one region after another. Frontiersmen (hunting was predominantly a male activity) 'mined' animal resources and fisheries as if they were mineral deposits of gold, silver or iron, working them until they were exhausted. For a time, at least until frontiers 'closed' and state control was extended, wildlife was an unregulated, open-access resource that was available to all users on a 'first come, first served' basis. The cultural systems of indigenous peoples that had previously determined access to hunting grounds broke down following European colonisation. To contemporary European market hunters, who generally did not look beyond maximising their short-term profits, it made sound economic sense to exploit what was a common resource until it was gone. In an increasingly competitive pursuit of resources, there was little incentive to conserve wildlife for the future.

Although approaches to commercial hunting varied under different colonial regimes around the globe, some common patterns and characteristics can be identified. The first European market hunters to penetrate new environments – from the Americas and southern Africa to the Arctic Ocean – tended to view the abundance of wildlife they encountered as a resource 'windfall', and every advance was accompanied by widespread asset-stripping of valuable mammal and fish stocks. Responding to metropolitan consumer demands for wildlife products, the recurrent pattern of commercialised hunting both on land and sea was broadly as follows: the exploitation of local populations of preferred species until exhausted; where feasible, the location and 'mining' of new prey species that offered a viable alternative in the same area; then a move outward to find new hunting grounds once these lucrative 'target species' had been sequentially depleted. Island species in particular were, and are, extremely vulnerable to intensive market hunting and other human activities. Today, disproportionate numbers of endangered or threatened species are to be found on islands. For example, the millions of sea turtles that nested on the beaches of Caribbean islands such as the Caymans – first named *Las Tortugas* (The Turtles) by Christopher Columbus in 1503 – had been hunted almost to extinction by 1900 for their flesh, eggs and shells. Turtle soup became a popular dish in imperial Europe, and turtle shells were used to make jewellery and ornaments. Only remnants of the Cayman Islands' formerly vast green sea turtle population, the most prized of the turtle species found in Caribbean waters, have survived.

The world hunt was important in altering relationships between indigenous peoples and their environments. It was also instrumental in weakening their societies and cultures. To meet rising consumer demands, European hunter-traders recruited large numbers of Native Americans, Africans and others to track, kill and process wild animals for export. These peoples were a cheap and mobile workforce who, for the most part, became willingly involved in commercial hunting to gain access to desirable European trade goods, such as metal tools, firearms, ammunition, textiles, tobacco and alcohol. Precious wildlife resources were extracted at minimal cost; although it should be noted that European merchandise had a high technological and symbolic value for indigenous peoples. Consequently, aboriginal hunting economies changed, with many becoming oriented primarily towards international markets rather than subsistence. Traditional restraints and taboos that had protected against overhunting and overfishing were often disregarded as indigenous peoples became 'enthusiastic consumers'. Unsustainable levels of market hunting undermined aboriginal subsistence systems, as wildlife resources critical to their well-being became scarce. Not only that, as communities became ever more reliant on imported goods acquired from European trading posts, many native crafts and skills disappeared – bow and arrow making, for example – threatening the survival of their cultures. A very similar pattern of subcontracting out the labour involved in commercial hunting to indigenous peoples, and of growing economic dependence as a result of trade, also emerged in Tokugawa Japan (1603–1868), following its expansion into Hokkaido.

If hunting for world markets had a corrosive effect on indigenous peoples' modes of life, it also had a devastating ecological impact, taking a heavy toll on biodiversity. From the late sixteenth century, an insatiable demand for furs in Europe and China – for warmth, fashion and status – saw the spread of commercial hunting across North America and Siberia. By this time, populations of European fur-bearing animals, such as bear, beaver, marten, otter and sable, were already in serious decline. Companies founded by the Dutch (Dutch West India Company, 1621), French (Company of New France, 1627), and English (Hudson's Bay Company, 1670) led the way in organising the North American fur trade, with cut-throat competition between them placing increasing pressure on animal stocks. With firearms and steel traps coming into common use from the seventeenth and eighteenth centuries respectively, hunting became more efficient too – compounding this pressure. Table 2.1 below shows the annual average yields for

Table 2.1 Furs harvested in North America (annual averages)

Species	*1700–63*	*1780–99*	*Percentage change*	*1830–49*	*Percentage change*
Beaver	179,268	263,976	47.3	77,654	−70.6
Racoon	91,637	225,115	145.7	322,759	43.4
Marten	51,315	88,856	73.2	130,283	46.6
Fox	18,411	20,360	10.6	79,056	288.3
Bear	16,033	26,833	67.4	13,229	−50.7
Mink	15,730	20,680	31.5	144,719	599.8
Otter	11,525	36,326	215.2	20,169	−44.5
Muskrat	10,432	177,736	1,603.8	849,865	378.2
Lynx/bobcat	10,179	17,277	69.7	35,443	105.1
Fisher	3,373	8,480	151.4	10,412	22.8
Wolf	1,830	16,461	799.5	8,899	−45.9
Wolverine	608	1,430	135.2	1,318	−7.8
Total	410,341	905,530	120.2	1,693,806	87.5

Source: John Richards *The Unending Frontier: An Environmental History of the Early Modern World* (Berkeley, Calif.: University of California Press, 2003). Reproduced with permission of University of California Press.

North American furs, harvested primarily for export to European furriers and hatmakers, spanning two centuries.

Such wholesale slaughter meant that fur-bearing animals were soon very scarce in eastern North America, forcing European hunter-traders and their indigenous partners – most notably the Iroquois – to push further and further west in search of more prey (which led to warfare and conflict at the advancing frontiers of empire over the control of new hunting grounds). Initially, the beaver was the main target species: hardest hit because its fur made sought-after, high-quality felt hats. The pre-colonial beaver population in North America has been conservatively estimated at about 50 million animals. But, as Table 2.1 helps to illustrate, its numbers had crashed by the mid-nineteenth century due to overhunting. By the early twentieth century only around 100,000 beaver remained, and it was locally extinct in many of its former ranges. Overall, the figures show that the fur trade decimated a broad spectrum of North American wildlife, with bears, otters, wolves and wolverines proving especially vulnerable to human predation.

Cold climates produced the most coveted and luxuriant furs for making warm garments. At the same time as the English, French and their allies were pressing westwards across North America, the Russians were expanding eastwards into Siberia in what has been described as one 'epic circumpolar quest for fur'. By 1800, the fur trade had come full circle as market hunters from two continents met on the Pacific coast of North America (the Russians were the first to arrive by sea in the 1730s). In contrast to the English and French, who traded with Native Americans for pelts, the Russians imposed the *iasak,* an annual tribute or tax, on the conquered Siberian peoples which had to be paid in furs – preferably sable. Throughout the seventeenth century, a combination of *iasak* collections and private hunting yielded an annual average of between 200,000 and 300,000 sable pelts, providing some 7 to 10 per cent of Russian state revenues. When supplies of sable began to dwindle – they could be found only in south-eastern Siberia by 1750 – the Russian state pragmatically accepted ermine, fox, marten, wolf and even squirrel pelts as payment of the *iasak.* However, unrelenting predation by both indigenous and Russian hunters was to rapidly strip these animals from the Siberian landscape. The opening up of rich hunting grounds along the coasts and islands of the northern Pacific Ocean gave frontiersmen access to new sources of fur, including large populations of the much-prized sea otter. In 1778, the best sea-otter pelts sold for the premium price of $120 in China. But in a highly competitive environment, with scant regard for the long-term conservation of resources, they too were soon hunted out. By the 1840s, despite the resilience of less valuable animals such as raccoons and muskrats, the circumpolar quest for fur was effectively over.

China had been the main marketplace for Russian hunter-traders after 1690, where they exchanged Siberian and Pacific coast furs for commodities such as gold, silver, silk, tea and porcelain. Similarly, until the nineteenth century, Chinese demand, together with consumerist pressure emanating from India and the Middle East, also propelled the lucrative international ivory trade. Elephant ivory was used for making ornamental statuettes, official seals, combs and jewellery, as well as for inlay work on expensive items of furniture. Africa was the world's foremost source of supply, and for centuries sub-Saharan ivory resources had been exploited by African peoples without any serious reduction of the elephant population. East African ivory, for example, had been exported via Muslim intermediaries to Asian and Arab markets since the Middle Ages. However, in the nineteenth century, the demand for ivory in Europe and the USA – to manufacture cutlery handles, billiard balls and piano keys – dramatically increased the volume of the trade. By the early 1880s, it has been estimated that an average of 12,000 elephants were being killed each year in East Africa alone. European hunter-traders were in the vanguard of exploration and expansion in Africa, lured ever deeper into the interior in search of ivory. As the hunting frontier moved outwards from the Cape Colony, elephants were cleared from much of southern Africa by the 1870s, and it was increasingly difficult to find them in Central and East Africa by the turn of the twentieth century. As in the North American fur trade, Europeans subcontracted out much of the hunting to indigenous peoples who, in return for firearms and other trade goods, probably killed more elephants on their behalf than they shot themselves. As elephants became scarce, commercial ivory hunters turned their attention towards the rhinoceros and hippopotamus. Rhinos were targeted for their horn (sold powdered in Asia as a medicinal 'cure-all'), and hippos for their teeth (used for making dentures as well as decorative objects), but these great pachyderms were also fast disappearing by 1900.

The European fashion industry's adoption of a traditional item of African dress – ostrich feathers – as a 'style accessory' contributed to setting off a craze for exotic plumage that was to wreak havoc on bird populations worldwide. Choice feathers had long been used by African, Amerindian, Asian and European rulers and elites, for ceremonial headdresses, feather standards and feather fans, as elaborate symbols of their power and position. In early modern Japan, for example, eagle feathers were a potent signifier of authority over others, and they were a valuable trade commodity. But it was in the second half of the nineteenth century that the plumage trade developed into a serious ecological threat, as feather boas and bird hats – which sported the heads and wings of birds as well as plumes – became a 'fashion fundamental' for women on both sides of the Atlantic. Ounce for ounce, the highest-quality ornamental feathers for feminine attire, such as ostrich plumes and *aigrettes* (the nuptial feathers of the egret), were worth more than gold. London was the centre of the international plumage trade, importing and re-exporting bird skins and feathers from the British Empire and elsewhere around the world. While statistical data on the scale and scope of the trade are few, Table 2.2 below shows the great mass of feathers that passed through British customs between 1895 and 1919.

In 1920, as trade networks disrupted by the First World War revived, critics of the plumage trade claimed that over 35 million bird skins were being annually imported into London for millinery purposes. Although this figure was undoubtedly exaggerated, by this time plumage hunters had pushed American egrets, West Indian hummingbirds, New Guinean birds of paradise and numerous other species to the verge of extinction (the South African ostrich had become domesticated). However, a fortuitous change in women's fashion during the early 1920s considerably reduced the demand for exotic feathers and allowed some wild-bird populations to recover from the damage caused by intensive market hunting.

While feathers fell in and out of fashion, the most prolonged assault on a specific species by commercial hunters was aimed at the whale. Before 1500, whaling was a relatively small-scale enterprise, with coastal communities such as the Inuit taking a small number of animals each year mainly to meet subsistence needs. Whales supplied meat and, more importantly, oil for heat and light, rendered from the blubber. The Basques pioneered deep-sea commercial whaling in the early sixteenth century, following their quarry across the Atlantic as far as the coasts of Newfoundland and Labrador, as local stocks of right whales in the Bay of Biscay had diminished due to overharvesting. Whalers first targeted right and bowhead whales because they were abundant, slow-moving and unaggressive, which made it possible to hunt them using a simple technology: hand-launched harpoons. Whaling took a similar form to the other key constituents of the world hunt: the exploration of unknown waters, the repeated discovery of rich whaling grounds and the rapid depletion of new stocks from the Arctic to the southern oceans. By the seventeenth century, other European nations, most notably the Dutch, Germans and English, had also established deep-sea fleets to hunt whales for their valuable oil and baleen (whalebone). Commonly utilised North Atlantic whaling grounds were already coming under heavy pressure by the mid-eighteenth century, prompting a move into the

Table 2.2 Imports of exotic ornamental feathers (including ostrich feathers) into the United Kingdom (quinquennial averages)

Dates	*Pounds weight*
1895–9	1,277,772
1900–4	1,401,122
1905–9	1,539,531
1910–14	1,166,706
1915–19	440,564

Source: R. J. Moore-Colyer, 'Feathered Women and Persecuted Birds', *Rural History* (vol. 11, April 2000, pp. 57–73). © Cambridge Journals, published by Cambridge University Press, reproduced with permission.

Pacific Ocean in the late 1780s as catches dropped to an unprofitable level. The Industrial Revolution gave fresh impetus to the worldwide search for stocks, as whale oil proved suitable for the lubrication of machinery and whalebone – the 'plastic' of the period – was used extensively in the manufacture of corsets, umbrellas and the like. But after around three centuries of relentless commercial hunting the most easily captured whales were gone. By 1900, whaling was in terminal decline as populations of right and bowhead whales collapsed and bigger, faster species such as Antarctic humpback and blue whales were too difficult to catch and process using existing methods. However, new technological advances made by the Norwegians, particularly the harpoon gun and the factory ship, were to give the industry a new lease of life.

In the early twentieth century, the Norwegians, having developed the technologies to catch rorquals (larger whales), opened up the last unexploited whaling grounds in the Antarctic seas. Demand for whale oil products remained high. As well as being an effective lubricant, whale oil was widely used to make margarine, soap and even explosives. Others soon adopted the harpoon gun and modern methods of processing, including new entrants into the pelagic whaling industry such as Argentina, Japan and Russia, which, as Table 2.3 demonstrates, impacted dramatically on whale populations around the world.

By the eve of the 1982 ban on commercial whaling, the largest and most valuable species – bowhead, right, humpback and blue whales – had already been hunted out. Whales reproduce slowly, and there is no guarantee that remnant populations will ever recover to their initial size. Today, only the smaller minke whale survives in sufficient numbers to be killed for scientific purposes (although most turn up for sale in Japanese seafood restaurants). Commercial hunting has played the dominant role in depleting whale

Table 2.3 Estimates of whale populations in the 1970s (in thousands of whales and percentages of initial populations).

Species	*Southern Hemisphere*	*North Pacific*	*North Atlantic*
Bowhead	Not present	1–2 (10%)	0.1 (1%)
Right	3–4 (10%)	0.4 (unknown)	0.2 (unknown)
Humpback	2–3 (3%)	2 (unknown)	1.5 (70%)
Blue	7–8 (4%)	2 (40%)	1 (10%)
Fin	80 (20%)	14–19 (40%)	31 (unknown)
Sei	50–5 (33%)	21–3 (50%)	2 (unknown)
Bryde's	10 (90%)	20–30 (100%)	Unknown
Minke	120–200 (80%)	Unknown	10 (unknown)

Source: adapted from Ray Hilborn, 'Marine Biota' in B. L. Turner, William C. Clark, Robert W. Kates, John F. Richards, Jessica T. Mathews and William B. Meyer (eds), *The Earth as Transformed by Human Action* (Cambridge: Cambridge University Press, 1990).

populations, ocean by ocean and species by species, over the past 500 years. As was the case with other open-access wildlife stocks, whalers usually took all that was available to them rather than leave behind valuable resources for the potential benefit of their rivals.

Settler hunting

To the first European settlers in lands as geographically distant and ecologically diverse as North America and South Africa, native wildlife represented both a windfall opportunity and a serious threat. Many enterprising agriculturalists subsidised their activities by hunting for the market, investing the proceeds of the trade in furs, feathers, ivory and other animal products into the farming economy. They used the money to buy seed, ploughs and livestock among other things. Moreover, hides, skins and pelts provided articles essential to settlers everywhere, such as harnesses for draught animals, wagon whips, blankets and even clothing for the poorest immigrants. But meat was perhaps the most important subsidy available to frontier farmers. They subsisted on the virtually cost-free meat of wild animals – venison and buffalo beef in South Africa, for example – while establishing their farms and smallholdings, thus avoiding the need to slaughter their own livestock. Along the expanding frontiers of empire, most settlers relied heavily on hunting wildlife for food until their farms became fully productive.

But once farms were profitable, with cash crops and stock-raising doing well, the wildlife that had acted as an indispensable 'life-support system' for the earliest settlers was increasingly seen as a threat. It was no longer in the farmers' interest to coexist with wild animals that damaged their crops and competed with their sheep and cattle for grazing. Frontier farmers hunted tirelessly to defend crops and extend pasturage. As the eradication of troublesome animals became a priority, shooting, trapping and poisoning became an everyday part of agricultural life. In North America, herds of deer, antelope and bison were either shot out or moved on. Australia's main grazing animals, kangaroos, were pushed out of eastern coastal lands by the spread of sheep and cattle, with both shepherds and stockmen making great efforts to exterminate them. European farmers in South Africa also targeted wildlife that threatened their livelihoods, with elephants, hippos, antelope and other large herbivores all killed in large numbers to protect cultivated areas and grazing lands. As wild animals retreated, they were replaced by livestock. In 1701, settler farmers in South Africa owned around 10,000 cattle (including oxen) and 53,000 sheep. There were more than 10 million sheep in the colony by 1875, including non-natives such as the wool-bearing merino (wool was South Africa's leading agricultural export at this time). The explosive growth of livestock populations, and the introduction of alien species by European farmers, also diminished biodiversity by crowding out native creatures.

To protect growing herds of cattle, sheep and other domestic stock, predators such as big cats, wolves and bears were exterminated as dangerous vermin. Because settler hunting and farming had made their natural prey scarce, many large carnivores took to killing livestock. Bounties began to be paid for wolf scalps in North America as early as 1630, a system that was extended across the colonies of European empires. In 1656, the Dutch East India Company offered bounties of six Spanish reals for killing a lion, and four for a leopard or hyena, to settlers in the Cape Colony. To make pastures safe for sheep in Australia after their introduction in 1792, bounties were soon paid on the scalps of the thylacine (the Tasmanian wolf or tiger) and the dingo. Indeed, there were bounties on a wide range of species, both carnivorous and herbivorous, that were perceived as a threat to the progress of agriculture. In 1926–7, for example, Australian farmers killed 48,951 emus, 40,944 crows, 7,093 scrub magpies and destroyed 45,456 emu eggs in Queensland alone. They pocketed bounties of over £9,500 for exterminating these birds, which disseminated the seeds of a highly invasive weed – the impenetrable prickly-pear cactus – throughout the state. The main targets for bounty hunters, however, were mammalian carnivores that preyed on livestock and, occasionally, people. As populations of large carnivores are usually very low in relation to those of their prey, the drive to control them as designated pests seriously depleted their numbers (although smaller, faster-breeding predators such as North American coyotes and Australian dingoes survived eradication campaigns).

At the same time, changing land use since European settlement – the conversion of forests to pasture and grassland to cultivated fields – destroyed or fragmented natural habitats, which caused further losses in faunal biodiversity. Settler hunting, then, severely reduced the numbers of large wild animals, while the bounty system (which persisted well into the twentieth century) sanctioned the total eradication of so-called 'vermin' in the agricultural and stock-raising areas of the 'neo-Europes'. In much of the Old World, especially western Europe and the Mediterranean region, the 'war against wild animals', as Mark Elvin has termed it, had already turned decisively in favour of farming communities by the early sixteenth century.

Sport hunting

Up until the nineteenth century, most of the hunting that accompanied European expansionism had been undertaken for utilitarian purposes: to satisfy market demands for wildlife products and to protect crops and livestock. But as empires were consolidated and extended during the Victorian era, the hunting of 'big game' – particularly in Africa and India – became a popular recreation of the colonial elite (military officers, senior government officials, large landowners and the like). The involvement of the European upper classes was to transform utilitarian hunting into the Hunt, which emphasised spectacle and political dominance of the imperial environment as well as 'sportsmanship': the ethos of fair play that came to govern this elite form of hunting. The elephant-borne tiger hunt in India, where the British Raj consciously adapted the methods of the former Mughal emperors, was undoubtedly the most ostentatious variant of the Hunt, and an effective demonstration of imperial power into the bargain. Hunting etiquette at such grand events required that the viceroy or governor bag the largest tiger, allowing British rulers to show their mettle. For example, Sir John Hewitt, Lieutenant-Governor of the United Provinces, shot at least 150 tigers before vacating the post in 1912. Moreover, as John Mackenzie has observed, many of the viceroys of India – Ripon, Dufferin, Curzon, Minto, Irwin, Willingdon and Linlithgow – seem to have been chosen as much for their sporting prowess as for their political accomplishments. Big-game hunting, which also included dangerous quarry such as panthers, wild boar, buffalo and bears, was an important symbolic means by which Britons could assert themselves as worthy successors to the Mughals.

By the late nineteenth century, sport hunting more generally was regarded as essential for building and testing 'character'. For the gentleman hunter, success was not measured simply by the size of the bag but by how the sporting encounter with wild animals was conducted. Unlike market and subsistence hunters, who aimed to dispatch wildlife with the bare minimum of time and effort, the true sportsman relished the 'thrill of the chase'. Sport hunting could be dangerous as well as exhilarating, and the 'manly' attributes it developed – courage, endurance and marksmanship – were valued highly at the turbulent frontiers of European empires, where conflict and warfare were endemic. To make the contest between humans and animals meaningful as firearms technology improved, hunting became hedged round with rules and standards that gave the latter a 'sporting chance'. The 'sportsman's code' discouraged improper methods of hunting, such as shooting indiscriminately into large herds, night-shooting using lights, or lying in wait for animals at water holes. It stressed the need to acquire knowledge about nature to identify and track prey, to avoid shooting female and immature animals, and to make a clean kill – a single shot to end the contest was the ideal. Campaigns to eradicate large carnivores continued unabated, however, because they competed for preferred game animals such as deer and antelope (which, by contrast, were to be shot only in moderate numbers). Gentlemen hunters displayed the heads and skins of their kills – big cats, deer, antelope and other wild animals – on the walls and floors of their residences as prized trophies. They were also employed to hunt on behalf of modern museums, such as the Natural History Museum in London and the Smithsonian Institution in Washington. These national institutions exhibited collections of stuffed creatures from around the globe for the 'rational amusement' of the general public, advertising human mastery over the animal kingdom.

Although most sport hunters were committed to fair play, the quantity of game they shot often ran to excess. In 1909, Theodore Roosevelt, former President of the USA and renowned sportsman, visited Kenya on a hunting trip sponsored by the Smithsonian. His expedition shot the massive total of 512 head of big game, far more than were needed for scientific research and museum displays. The competitive rivalry between sport hunters – the desire to 'bag' the largest specimen or to shoot the greatest number of game – could result in the extravagant slaughter of wildlife. But as hunting for pleasure gained in popularity among the upper classes during the second half of the nineteenth century, and as game stocks dwindled in imperial environments, the need to preserve wildlife for 'sport' became increasingly apparent. Colonial regimes attempted to place limits on the size of bags, to shorten the open season and to protect females and young animals. They even banned outright the hunting of some rare species. The hunting methods of indigenous peoples and poor immigrants – the use of snares, pitfalls, traps, nets and poison – were condemned as cruel, wasteful and 'unsporting'. New game regulations, varied as they were from place to place, sought to restrict social access to hunting by prohibiting the use of such inhumane techniques. The settler right to hunt as one pleased was curtailed, and in some cases indigenous peoples were wholly excluded from hunting wildlife. It was argued that their standards of living would never improve while they relied on subsistence and market hunting and they were able to avoid regular wage labour. Towards the end of the nineteenth century, harking back to the traditional European model that reserved game for the wealthy, upper-class sportsmen tried to secure a monopoly over diminishing wildlife resources by successfully lobbying for the introduction of colonial legislation that restrained lower-class and 'native' hunting.

By the early years of the twentieth century, sportsmen's concerns over disappearing wildlife – especially in Africa – led to the first international agreements to control hunting. In 1900, several European imperial powers signed the Convention for the Preservation of Wild Animals, Birds and Fish in Africa, a German and British initiative that encouraged all participants to enact and enforce game legislation and that advocated the creation of game reserves and hunting only by licence. These aims were extended in 1933, when the Convention on Preservation of Fauna and Flora in their Natural State was signed, which called upon contracting governments to establish national parks and 'strict natural reserves' – areas that were off-limits to hunters – to protect endangered animals and their habitats in Africa and other parts of the world. Belgium, Britain, Egypt, France, Italy, Portugal, South Africa, Spain and the Anglo-Egyptian Sudan were represented at the meeting (although not all implemented the convention's provisions fully in their territories). In the USA, wildlife had found refuge in national parks and game reserves since the late nineteenth century, but the Washington Convention on Nature Protection and Wild Life Preservation in the Western

Hemisphere (1940) approved the creation of 'national parks, national reserves, nature monuments, and strict wilderness reserves' throughout the Americas. A supranational organisation, the International Union for the Preservation of Nature (later renamed the International Union for the Conservation of Nature and Natural Resources [IUCN]), encouraged the protection of species worldwide. Founded in 1948 by the United Nations Educational, Scientific and Cultural Organization, the Swiss League for Nature Protection and the French Government, the IUCN promoted the national-park idea and compiled extensive datasets on endangered species – its well-publicised 'Red Lists'. Its biggest achievement to date has been the Convention on International Trade in Endangered Species of Wild Fauna and Flora, or CITES (1975), which currently involves 167 countries in protecting over 30,000 species from trade that threatens their survival.

What began as a policy to preserve elite access to game was to evolve into a worldwide framework for wildlife conservation after the Second World War. More than 10 per cent of the earth – including some ocean areas – has now been set aside as wildlife sanctuaries and bioreserves, where creatures are shielded from disruptive human activities. Today, people are welcome only as camera-toting tourists (indigenous peoples were often removed from their lands so that parks and reserves could be established). However, critics point out that this global patchwork of national parks and bioreserves is essentially an exercise in the human selection of species: we decide on the endangered animals that are to be protected, in what numbers and in which locations.

Case study: the destruction of the American bison

The destruction of the American bison is one of the most important and widely debated episodes in the history of the world hunt. How many bison ranged across the Great Plains before Euro-American expansion and settlement has been a major bone of contention. Early scholarly estimates, based on the observations of nineteenth-century explorers such as Meriwether Lewis and William Clark, suggested that there were somewhere between 75 and 100 million bison inhabiting the grasslands between Canada and west Texas. Recent research, however, by taking into account the range-carrying capacity, has lowered the figures considerably, indicating that the Great Plains supported no more than 30 million bison prior to the arrival of the Europeans in North America. In addition, these studies have shown that bison populations were extremely volatile, fluctuating constantly due to the effects of drought, grassfires, blizzards and wolf predation. Nonetheless, until the nineteenth century, the immensity of American bison herds, in the words of the contemporary artist-naturalist John James Audubon, was 'impossible to describe or even conceive'.

Whether Native Americans were really the 'first ecologists' is a question that has also divided historians concerned with bison-hunting. Most once assumed that Native Americans lived in relative harmony with their environments and that their cultural values and hunting practices were designed to limit demands on natural resources. Plains Indians – Arapahos, Assiniboines, Blackfeet, Cheyennes, Commanches, Crow, Kiowas, Sioux and others – only harvested around half a million bison every year for domestic consumption (a sustainable figure given that it was such a prolific species). For them, it served as a 'tribal department store'. Almost the whole of the animal had value, including: flesh, tongues, most organs and bone marrow for food; hides (hair off) for tepee covers, shields, moccasins, leggings and other clothing; robes (hair on) for winter clothing, bedding, gloves and ceremonial costumes; sinew for thread, bowstrings and snowshoe webbing; dung for fuel; horns for arrow points, bow parts, cups and containers; bones for tools and teeth for ornaments. This is far from being an exhaustive list. The Blackfeet, for example, obtained more than 100 specific items that sustained daily life from the bison. But in emphasising Native American ingenuity in turning a variety of bison parts into useful products, considerable evidence of waste and overkill tended to be overlooked.

In recent years, the role played by Native Americans in the near extinction of the bison has been more closely studied. Archaeological evidence has revealed that traditional methods of hunting, such as stampeding bison off cliffs or into box canyons, could be remarkably wasteful. For instance, the Olsen-Chubbock 'buffalo jump' site in southern Colorado contained the bones of some 200 bison (males and females, adults and juveniles), but over forty animals at the bottom of the pile had either not been touched or only partly butchered. By the late eighteenth century, the adoption of the horse and the rifle had made Plains Indians more effective and selective hunters. They preferred to kill bison cows, because their meat was more succulent and their skins were easier to work. After 1700, almost all Plains Indians had become enmeshed in commercialised hunting, exchanging bison tongues, hides and robes for European trade goods and horses. Native Americans began to slaughter the bison for their tongues and skins alone, leaving the rest of the carcass to rot. Between the 1830s and the 1860s, as demand increased, they supplied firms such as John Jacob Astor's American Fur Company with over 100,000 bison robes annually. In combination with environmental factors, especially drought, the Plains Indians' preference for killing females, the competition for grazing from their horses, and the pressures of both subsistence and market hunting began to speed the decline in bison numbers.

Contrary to the old orthodoxy, there were elements in Native American cosmologies that were also destructive to the bison. The Plains Indians believed that when the great bison herds dispersed during the winter months they went to grasslands underground, reemerging every spring in inexhaustible swarms 'like bees from a hive'. If bison herds did not reappear in expected numbers, it was because many had not yet left their subterranean prairies. Such a notion undoubtedly discouraged the conservation of what was a diminishing resource, as Plains people could still hunt prodigiously secure in the knowledge that there were always more animals grazing down below. Today, few historians would accept uncritically the idealised image of the 'ecological Indian', who consciously pursued a lifestyle that left few traces on the land. But the pressures which Plains Indians exerted on the bison were slight in comparison with Euro-American market hunters.

The bison population collapsed rapidly from the mid-nineteenth century. The demand for bison hides for the tanning industry was the primary cause. In 1850, tanning was the fifth-largest industry in the USA, and it grew in value by supplying markets at home and overseas with a variety of leather products – most importantly the heavy belting that powered machinery in mills and factories. Demand outstripped the supply of cowhides in the USA, forcing tanners to import them from Argentina and elsewhere in Latin America. But in 1870 a process for transforming bison skin into tough, elastic leather was discovered by tanners in Philadelphia: a material that was ideal for use as industrial belting. Thousands of Euro-American commercial hunters – armed with powerful rifles developed during the Civil War – flooded onto the Great Plains, killing around 2 million bison annually for their hides between 1870 and the early 1880s. At the same time, the extension of the railroads to the plains made transporting bison hides to tanneries in the East both quick and inexpensive. Trains also brought sportsmen who were keen to 'bag a buffalo', but it was Euro-American market hunters that almost annihilated the bison. In 1876, a single hide-hunter shot nearly 6,000 bison in the space of just two months (although market hunters usually killed between 2,000 and 3,000 buffalo per season). As bison were a common resource, it made commercial sense for hunters to take all that was available to them. After little more than a decade of industrial-scale slaughter, the last shipment of hides from the Plains took place in 1884. By the end of the nineteenth century, the American bison population numbered fewer than 1,000 animals.

The last remnants of the great bison herds found refuge in Yellowstone National Park and in city zoos, far from their former rangelands. Their destruction saw the reshaping of the Great Plains into cereal farms and pastures, with millions of cattle and sheep replacing them on the grasslands – preventing any long-term recovery of the bison to its former numbers. The extension of the national park and reserve system in Canada and the USA during the first three decades of the twentieth century, largely at the behest of sportsmen and nostalgic Eastern philanthropists, did help to save the bison from extinction – but at a cost. It became an 'imprisoned' and semi-domesticated species. Bison refuges, such as Yellowstone and Wood Buffalo National Park in Canada, were soon forced to cull their populations, as without predation they quickly rebounded. Although this practice was curtailed in the late 1960s, any bison straying beyond park boundaries in search of grazing still risks being shot by ranchers worried about the spread of bovine diseases. In 2000, there were nearly 250,000 bison in North America. But the growth of intensive farming, cities and suburbs, shopping centres and industrial parks, and rail and road networks – allied to the fear that bison carry bovine tuberculosis – means that it is highly unlikely that they will be allowed to disperse widely across the landscape once again.

Conclusion

Over the past 500 years, the activities of commercial, settler and sport hunters have dramatically reduced wildlife populations around the globe, especially the numbers of 'charismatic megafauna'. Many species were lost completely – the thylacine, blaaubok (blue antelope), passenger pigeon and numerous others – particularly during the nineteenth and twentieth centuries, when the pace of economic activity accelerated. As technologies advanced with the Industrial Revolution, and the consumer demands of fast-growing human populations increased, the pressure on wildlife intensified. At the expanding economic and colonial frontiers of empire, indigenous cultural systems that had controlled local access to wildlife resources broke down. Instead, access to wildlife became a competitive 'free-for-all', substantially depleting common resources which, to the first European frontiersmen, appeared to be inexhaustible. Indigenous peoples were not 'ecological saints' prior to the arrival of Europeans. While they lived lightly on the land, pre-colonial hunter-gatherer societies often failed to husband natural resources wisely. However, the commercialisation of indigenous cultures, usually on inequitable terms, and integration into world markets altered long-established hunting patterns and multiplied the numbers of wild animals they killed. Native Americans, for instance, who had coexisted with the bison for some nine millennia without seriously overexploiting them, participated in the commercial hunt for robes and hides until 'the tail of the last buffalo' vanished.

As the world hunt rapidly eliminated wildlife from natural ecosystems, imported species such as cattle and sheep replaced them, modifying their habitats and making it difficult for remnant populations to recover. The total populations of cattle and sheep worldwide were estimated at 1.3 billion and 1.2 billion respectively in 1990. In supplanting native species, European domesticated animals – rather than adding to biological diversity – simplified ecosystems on a major scale in the Americas, Australia, New Zealand, southern Africa and elsewhere. Habitat destruction and introduced species, by leaving little space for native wildlife, were also key factors in reducing terrestrial biodiversity. On the world's oceans, however, overharvesting was unquestionably the main cause of declining whale and fish populations. From an ecological perspective, the long-term effects of oversimplifying ecosystems by depleting the numbers of keystone species such as whales, bison, elephants and tigers are still not entirely clear. But, to quote Edward O. Wilson, 'it is reckless to suppose that biodiversity can be diminished indefinitely without threatening humanity itself.'

In the last decades of the nineteenth century, as it became apparent that many species of wildlife were in drastic decline, public interest in their conservation began to increase worldwide. Initially, this movement was driven mainly by European and American sportsmen. They founded bodies such as the Boone and Crockett Club (1887) and the Society for the Preservation of the Wild Fauna of Empire (1907) to campaign for the creation of game reserves and the passage of legislation that would restrict the 'unsporting' hunting activities of indigenous peoples and poor immigrants (who were blamed for overexploiting wildlife stocks for base commercial gain). Their legacy has been the establishment of national parks and bioreserves around the world and a series of international treaties and conventions such as CITES, designed to protect endangered species. But present-day critics of what is still a Western-led and -dominated conservation movement have complained of a 'green imperialism' that is little different from that of an earlier age, as customary hunting practices are outlawed and the poor inhabitants of newly designated parks are removed to keep them 'free of human interference'. Even so, international agreements to protect wildlife and their habitats are very difficult to enforce in remote regions of many countries. Illegal hunting remains a leading cause of species endangerment and extinction. According to the World Wildlife Fund, the global conservation organisation, the profits from the illicit trade in elephant ivory, rhino horns, big cat pelts and other animal products are currently worth at least $15 billion annually. Without better environmental management, involving indigenous and local communities more closely in decision-making, it is estimated that 25 per cent of the earth's mammals and 12 per cent of its birds may disappear over the next century.

Further reading

David Arnold, *The Problem of Nature: Environment, Culture and European Expansion* (Oxford: Blackwell, 1996).

William Beinart and Peter Coates, *Environment and History: The Taming of Nature in the USA and South Africa* (London: Routledge, 1995).

Thomas R. Dunlap, *Nature and the English Diaspora: Environment and History in the United States, Canada, Australia and New Zealand* (New York: Cambridge University Press, 1999).

Mark Elvin, *The Retreat of the Elephants: An Environmental History of China* (New Haven, Conn.: Yale University Press, 2004).

Alf Hornberg, J. R. McNeill and Joan Martinez-Alier (eds), *Rethinking Environmental History: World-System History and Global Environmental Change* (Lanham, Md.: Altamira, 2007).

J. Donald Hughes (ed.), *The Face of the Earth: Environment and World History* (Armonk, NY: M. E. Sharpe, 2000).

Andrew C. Isenberg, *The Destruction of the Bison: An Environmental History, 1750–1920* (New York: Cambridge University Press, 2000).

Shephard Krech, *The Ecological Indian: Myth and History* (New York: Norton, 1999).

Shephard Krech, John McNeill and Carolyn Merchant (eds), *Encyclopedia of World Environmental History* (New York: Routledge, 2004).

John McNeill, *Something New under the Sun: An Environmental History of the Twentieth Century* (London: Allen Lane, 2000).

John Mackenzie, *The Empire of Nature: Hunting, Conservation and British Imperialism* (Manchester: Manchester University Press, 1988).

Stephen N. Meyer, *The End of the Wild* (Cambridge, Mass.: MIT Press, 2006).

R. J. Moore-Colyer, 'Feathered Women and Persecuted Birds', *Rural History* (vol. 11, April 2000, pp. 57–73).

Rob Nixon, *Dreambirds: The Strange History of the Ostrich in Fashion, Food and Fortune* (New York: Picador, 2001).

Clive Ponting, *A New Green History of the World: The Environment and the Collapse of Great Civilisations* (London: Vintage Books, 2007).

John F. Richards, *The Unending Frontier: An Environmental History of the Early Modern World* (Berkeley, Calif.: University of California, 2003).

Joseph Sramek, '"Face Him Like a Briton": Tiger Hunting, Imperialism, and British Masculinity in Colonial India, 1800–75', *Victorian Studies* (vol. 48, summer 2006, pp. 659–80).

Ted Steinberg, *Down to Earth: Nature's Role in American History* (New York: Oxford University Press, 2002).

B. L. Turner, William C. Clark, Robert W. Kates, John F. Richards, Jessica T. Mathews and William B. Meyer (eds), *The Earth as Transformed by Human Action* (Cambridge: Cambridge University Press, 1990).

Brett L. Walker, *The Conquest of Ainu Lands: Ecology and Culture in Japanese Expansion, 1590–1800* (Berkeley, Calif.: University of California, 2006).

Edward O. Wilson, *The Diversity of Life (*Harmondsworth: Penguin, 2001).

Eric R. Wolf, *Europe and the People without History (*Berkeley, Calif.: University of California, 1997).

Chapter 3

Forests and forestry

The clearing of the world's forests is perhaps the most dramatic transformation of the earth's land surface by humankind. About 8,000 years ago forests covered almost half of the planet, a total area of more than 6 billion hectares. Since this time, the world's forests have reduced in size somewhere between 15 to 45 per cent. Estimates vary markedly as historical data on forest conditions, particularly for many developing countries, are unreliable, incomplete or not comparable. Allowing for the limitations of national data sources, the landmark Millennium Ecosystem Assessment (2005) – using supplementary information generated by new remote-sensing satellite technologies – nonetheless reported with 'high certainty' that some 40 per cent of the world's forests have been lost. While humans had slowly modified forested environments using fire and axe for millennia, the process began to accelerate after 1492 as European settler societies cleared forests in the Americas, southern Africa, Australia and New Zealand to create pasture and farmland. Most deforestation, however, occurred during the past two centuries with the introduction of mechanised methods of harvesting and transporting timber following the Industrial Revolution.

The development of the modern world economy – with Europe at the heart of international trade – saw a global decline in the area and quality of the three major forest types: boreal (northern), temperate and tropical. The great boreal forests are circumpolar, primarily composed of evergreen conifers such as pine, spruce and fir, and they stretch in a broad belt across Alaska, Canada, Scandinavia and Russia (the Siberian boreal forest being the largest). At the southern edges of their range, the boreal forests mix with deciduous (leaf-shedding) temperate forests. These forests include broad-leaved species such as oak, beech, elm, maple and willow, and were originally widely spread throughout North America, Europe and north-east Asia. Tropical forests are characterised by their astounding biodiversity. Mainly distributed in South America, west and central Africa, south-east Asia, Australia and the Indian subcontinent, they often contain over 100 tree species per hectare; far more than boreal and temperate forests. Many of these tree species have a high commercial value, such as teak, ebony, mahogany and sandalwood. Until recently, deforestation had been most intensive in the temperate world. Europe, for example, has virtually no primary (naturally regenerated) forest cover remaining. Since 1950, however, the focus has shifted to tropical forests, which are now being lost at a rate of over 10 million hectares – an area larger than Greece – per annum.

Deforestation has thoroughly transformed the natural environment, both in the temperate North and the tropical South, affecting the quality of human life and undermining the health of the planet. This chapter will explore the complex set of economic, socio-cultural, technological and natural forces that have caused forests to decline. It will also chart the growth and development of strategies to manage and conserve the world's forests. Environmental historians have shown a particular interest in the evolution of European 'scientific forestry' from the mid-eighteenth century onwards, arguing that it played a major role in the emergence of modern environmentalism. During the Enlightenment era, the idea of 'improving' forests and using their resources more rationally gained ground. But we will begin by looking at the importance of forests for people and for sustaining a habitable planet.

The importance of forests

Forested lands have always been utilised to meet a wide range of basic human needs, from food and fuel through to shelter and the spiritual. In the past, as the professor of forestry Donald W. Floyd neatly put it, 'human societies were literally built on wood.' Timber was so important for construction, shipbuilding and industry that, despite its relatively low economic value, it was transported in great quantities over long distances. From the fourth century BCE, far-ranging timber trade routes criss-crossed the Mediterranean basin to supply the material needs of the classical civilisations of Greece and Rome. By early modern times, timber was becoming a key commodity in world trade, 'transcending the common-sense rule' that only expensive goods such as sugar, silk, tea, coffee and tobacco should be shipped overseas. In the seventeenth century, for example, New England oak was exported to the timber-starved islands of Madeira (for wine casks) and the Caribbean (for sugar and molasses casks). European demands for masts, ships' timbers and naval stores (turpentine, pitch and tar), as well as lumber for general building purposes (from floorboards to roofing joists), were largely met by Baltic and Scandinavian forests, although imports of wood products from North America increased after 1700. Tropical hardwoods such as teak, ebony and mahogany were also traded internationally but to a much lesser extent, initially for use in the production of luxury items such as fine furniture, carved ornaments and musical instruments. By the early nineteenth century, however, the teak forests of Burma and India's Malabar coast were crucial to the British – the leading organisers and carriers of the global timber trade – for shipbuilding in the colonial port cities of Bombay and Calcutta. Indeed, forest resources were fundamental to the expansion of Europe's great imperial powers, as before 1850 most ocean-going ships – the mainstays of world trade and exploration – were constructed from wood.

Early industrial development was underpinned by wood, particularly in the USA, Germany and Japan. Until the nineteenth century, the energy to smelt metals, to fire bricks and ceramics, to make glass, to process salt, to refine sugar, and for many other forms of industry, came primarily from forests and woodlands. Vast quantities of wood were required to fuel traditional manufacturing technologies, especially for iron-making. Charcoal-fuelled furnaces produced tough and malleable iron that could be fashioned into everything from agricultural tools to weapons of war. To take but one example, in 1810 the USA used about 2 million tonnes of wood – the equivalent of 2,600 square kilometres of forest – to produce just 49,000 tonnes of pig iron. As late as 1856, over 75 per cent of American iron was still made using charcoal. This was because charcoal-smelted iron was considered superior in quality to that produced by coke-fuelled furnaces and because wood was so abundant in the USA. The mining of ironstone and other metal ores relied in no small part on forest resources as timber was widely used for pit-props. By the early twentieth century, Japan's mining industry was the second-largest consumer of the country's lumber after housing construction. Building railways – the epitome of industrial progress – also depended on access to immense quantities of timber for sleepers, water towers, bridges, telegraph poles and fuel (where coal was scarce), as networks rapidly expanded worldwide after 1840. By 1913, world railways totalled 1.1 million kilometres of track. Even after the arrival of mature industrialism, when the rise in importance of fossil fuels, iron and steel and the introduction of new materials such as concrete and plastics meant that societies and their economies no longer relied so heavily on wood, new uses for paper products, packaging and plywood and the continued utility of timber for construction made it an indispensable natural resource.

The emphasis on forest resources as essential to imperial and industrial expansion, however, must not overshadow their numerous other uses. Forests acted as a buffer against hunger for large numbers of the world's poorest people by providing fruits, nuts, honey and other foodstuffs to supplement inadequate diets. They furnished medicinal plants that could be used by indigenous healers to cure ailments and to improve health. Rural communities hunted for meat and furs in forested areas (see Chapter 2) and often used them as grazing grounds for their livestock. Forests also provided fuelwood, an everyday necessity for cooking and home heating. The diaries of pioneer settlers in North America show that as much as one quarter of their time could be spent in gathering, splitting and stacking firewood, which was vital for their survival during the intensely cold winter months. Globally, around 2 billion people in developing countries still depend on wood to meet their energy needs. At the turn of the millennium, fuelwood, mainly utilised for domestic purposes, accounted for 58 per cent of all the energy used in Africa, 15 per cent in Latin America and 11 per cent in Asia. Fuelwood currently makes up around 80 per cent of all wood consumption in the world's least developed countries and – used wisely – it can be a potentially renewable source of energy.

Forests have long been valued for the spiritual and recreational roles they play in many societies. In *The Golden Bough* (1890), the Scottish anthropologist James Frazer observed that the worship of trees was one of the oldest features of religious systems in almost every region of the globe. Sacred groves – stands of forest or woodland protected to please particular deities – were once common in ancient Greece, Rome and pre-Columbian America. They were regarded with reverence as the shelters of local gods, spirits and supernatural powers that might bring sun, rain and fertility – or ruin. Few people disturbed sacred groves for fear of these deities' retribution. They still persist in many parts of the world, in India, south-east Asia, Oceania, sub-Saharan Africa and isolated areas of the former Soviet Union. By the late nineteenth century, large tracts of forest also began to be set aside as recreational resources for European and American tourists. Wealthy city-dwellers, inspired by the work of Romantic and transcendentalist writers, artists and poets such as William Wordsworth, Jean-Jacques Rousseau, Caspar David Friedrich, Thomas Cole, Henry David Thoreau and John Muir, increasingly sought spiritual enrichment in 'wilderness' areas. The beauty and tranquillity of many forest environments made them desirable tourist destinations for those with the means to escape the hustle and bustle of urban life. From the establishment of the first national parks in the USA – Yellowstone (1872), Yosemite (1890) and Sequoia (1890) – to the growth of modern ecotourism in the Amazon region, forest-related recreation has provided a 'tonic for the jaded human spirit'.

The world's forests and woodlands have always provided – and continue to provide – 'ecological services' that are fundamental to the maintenance of a habitable planet. They are an important storehouse of biodiversity, containing at least two-thirds of the earth's terrestrial species (with the greatest biotic diversity occurring in tropical forests). But the International Union for the Conservation of Nature and Natural Resources estimates that 87 per cent of the world's reptiles, 75 per cent of its mammals, 57 per cent of its amphibians, 44 per cent of its birds and 12 per cent of its plants are now threatened by forest decline. Some 8,000 tree species (around 10 per cent of the total known to science) are currently facing extinction. In many regions, forests play an important role in stabilising natural landscapes by preventing soil erosion and reducing the risk of disastrous flooding and landslides. They help to protect water supplies, with more than 75 per cent of the world's accessible freshwater coming from forested catchments. Forests also contribute to the stability of the global hydrological cycle (the means by which water is circulated in the biosphere) by returning moisture back to the atmosphere as evapo-transpiration from the leaves of trees. Scientific studies show that forests are effective natural mechanisms for capturing and storing carbon dioxide, helping to regulate the global climate system and to mitigate climate change. It is clear that the value of the various 'ecological services' that forests provide – habitat for wildlife, soil and water protection, carbon storage and climate regulation – is high, despite not generally being recognised in financial terms either in the past or present. And the decline of forests threatens all of these essential functions at all levels: local, regional and global.

The drivers of deforestation

Concerns about deforestation have a long history. During the classical Hellenistic and Roman periods (1100 BCE to 565 CE), the ancient forests around the Mediterranean basin were heavily exploited for construction materials, ships' timbers and fuelwood for both domestic and industrial use. For the first time, writers such as Plato, Theophrastus, Strabo and Cicero provided detailed accounts of the causes of deforestation, as expanding city-states increased their consumption of wood products and cleared land to grow food. In his book *Critias* (c. 360 BCE), for example, Plato drew attention to the deforestation of Attica for essential wood supplies. Where once there had been 'an abundance of wood in the mountains', he lamented that only 'the mere skeleton of the land' remained. The patterns of resource use first described by Plato and other classical writers brought lasting change to the Mediterranean landscape, and they would persist until the industrial era. The history of deforestation in ancient Greece and Rome can be seen as the prologue to the modern onslaught on the world's forests, outlined in Table 3.1 below.

Table 3.1 Estimated forest clearances (× 000 km^2)

Region or Country	*Pre-1650*	*1650–1749*	*1750–1849*	*1850–1978*	*Total*
North America	6	80	380	641	1,107
Central America	18	30	40	200	288
Latin America	18	100	170	637	925
Oceania	6	6	6	362	380
USSR	70	180	270	575	1,095
Europe	204	66	186	81	537
Asia	974	216	606	1,220	3,016
Africa	226	80	42	469	817
Grand Total	1,522	758	1,700	4,185	8,165

Source: adapted from Michael Williams, 'Forests' in B. L. Turner, William C. Clark, Robert W. Kates, John F. Richards, Jessica T. Mathews and William B. Meyer (eds), *The Earth as Transformed by Human Action* (Cambridge: Cambridge University Press, 1990).

While the figures in Table 3.1 must be treated with caution (there are some rather conservative 'guesstimates'), they nevertheless provide a useful overall picture of the pace and scale of human-induced deforestation from before 1650 to the late twentieth century. It is estimated that over 8 million square kilometres of forests were cleared in eight major regions and countries of the world. Pre-1650, the three oldest-settled continents (Europe, Asia and Africa) show the highest rates of deforestation. Only in Europe did clearing slow down significantly after 1850, as concern over potential wood shortages and the rise of the Romantic Movement produced a new appreciation of forest landscapes.

Although rates of deforestation varied markedly, it is possible to identify the key factors that directly and indirectly affected forest cover globally. The main direct drivers of deforestation were, and still are, wood extraction and agricultural expansion. From the late fifteenth century, the emergence of a capitalist world economy – with western Europe providing the commercial dynamism that powered the system – placed increasing pressures on forests everywhere. As was the case with wildlife (see Chapter 2), the process of integration into international markets often disrupted or even destroyed the management systems of indigenous peoples that had previously regulated access to forest resources and curbed environmental degradation. As European empires expanded worldwide, at their frontiers timber resources began to be 'mined' vigorously for shipbuilding, fuelwood, housing construction and other purposes. Europeans had profligately cleared their own forests during the Middle Ages, particularly in the British Isles, Holland, Spain and Portugal (where just 5 to 10 per cent of their lands remained forested). In the New World, as David Arnold has described, Europe established 'a seemingly ceaseless sawmill able to keep its ships afloat and its drawing rooms supplied with elegant tables, chairs and writing desks'. Given its importance to the construction and maintenance of navies and merchant marines, timber was seen by the Europeans as a key strategic resource – much like oil is today – to be secured if necessary through warfare. The British, for example, fought three wars in quick succession with Burma (in 1824–6, 1852 and 1885–6) to win unrestricted access to its plentiful teak forests. But until the mid-nineteenth century the general European attitude was to extract wood as if it were an inexhaustible resource, with very little regard being shown for forest conservation.

Forests were also felled extensively to create another valuable resource: farmland. The primary driver of global deforestation was agricultural clearing for crops and pasture. After 1682, 'huge cropland expansion' saw the disappearance of mixed and coniferous forests in the growing Russian empire. As settler societies were permanently established in the temperate 'neo-European' lands of North America, southern Africa, Australia and New Zealand from the early modern period onwards, they too wasted no time in removing forests to set up farms. Pioneer settlers believed that tree size was a strong indicator of soil fertility. The larger the trees, the sooner they were cleared to grow food and raise livestock. North America provides the best documented and most dramatic example of neo-European deforestation. After 1607, pioneer farmers – with some assistance from lumbermen – cleared vast areas of forestland in US states stretching from Florida to Maine and in the Canadian Maritimes, Quebec and Ontario. By the early twentieth century, more than half of North America's eastern forests had been replaced by farmland that produced European crops such as wheat, oats and rye and profitable 'crops on hooves' too – European cattle, sheep and pigs – remaking the New World landscape into a replica of the Old. Farm-making activity in southern Africa, Australia and New Zealand had similar results. In Australia, for example, after 1788, trees were cleared to feed a rapidly growing settler population and to provide grassland for sheep and cattle (wool became the country's main agricultural export). The latest research shows that almost 400,000 square kilometres of Australia's south-eastern forests had disappeared by the early twentieth century, largely due to the expansion of agriculture. The data in Table 3.1 leave little doubt that the traditional cultures of Native Americans and Indigenous Australians (Aboriginal people), who for millennia had used controlled burning to alter forestland to meet their subsistence needs, were more sustainable.

As well as causing widespread deforestation in temperate 'neo-European' regions, agricultural expansion played a significant role in accelerating forest decline in colonised areas of the tropical world. After 1492, as European consumers developed an insatiable craving for sugar, tea, coffee, chocolate and tobacco – the spread of a 'soft drug culture' in the West – subtropical and tropical forests were cleared to make way for colonial plantations. Beginning with sugarcane cultivation on a modest scale in Madeira and the Canary Islands, the plantation system first developed by Iberian colonisers became a model for production that was widely imitated throughout the tropics. As the Dutch, British, French and North Americans developed plantation enterprises of their own in the Caribbean, South America, west Africa, south-east Asia, India and elsewhere, forest cover was systematically replaced by monocultures of sugarcane, cacao and tobacco plants and coffee and tea bushes. Although heavy taxation limited the consumption of 'luxuries' until the mid-nineteenth century, the advent of free trade reduced their price and helped to make so-called 'drug foods' a staple part of Western diets. In addition, cotton, rubber and other cash crops for industrial use were grown on colonial plantations, which also pulled tropical areas into the burgeoning international market economy. The human costs of the coercive plantation system, largely based on slave and indentured labour, have been well documented by historians. But the environmental degradation that went hand in hand with social inequality has tended to be overlooked. Plantation agriculture had a devastating impact on the forests of producing nations with Brazil, for example, losing over half of its subtropical Atlantic coastal forest (although the gold- and diamond-mining industries played a major role here too). Moreover, unequal and environmentally destructive trading relationships survived European decolonisation. Since 1950, more than 5 million square kilometres of tropical forests have been lost, in no small part to produce agricultural exports such as coffee, beef and biofuels (derived from crops such as sugarcane and soybeans), as well as construction timber and pulpwood, for consumption in the developed world.

However, societies that had largely resisted European intrusion also over-exploited their forests. In early modern Japan, a 'vast construction boom', presided over by a powerful ruling elite (*daimyo*), saw the severe depletion of its forest resources. The *daimyo* built great castles, mansions, Buddhist temples, Shinto shrines and new towns (over 200 were established between 1467 and 1590). As the Japanese preferred to build with wood rather than stone, these monumental construction projects required immense quantities of timber – decimating the country's forests from Kyushu to northern Honshu. Reconstructing vast fire-prone cities such as Edo (around 1 million inhabitants by 1720), which were repeatedly destroyed by conflagrations, also contributed to nationwide deforestation. Other significant causes of forest depletion in Japan included conversion to agriculture, demand for fuelwood, iron-making for the manufacture of weapons for samurai armies and the construction of ocean-going warships. By the late seventeenth century, wholesale logging had left Japan's three main islands largely devoid of old-growth forest, causing the Confucian scholar Kumazawa Banzan to complain that 'eight out of ten mountains of the realm have been denuded.' Japan's forests also came under great pressure in the Meiji period (1868–1912), as early industrialisation relied heavily on wood for fuel, iron-making and even machine-building. Japanese carpenter-mechanics built countless hybrid machines – such as power-looms used to manufacture textiles – that contained both wooden and metal parts. Wood played a central role in the nation's architectural tradition and in its modernisation, and its continued importance for house-building today has helped to make Japan the world's biggest timber importer.

The forest history of China is more difficult to reconstruct because little evidence – written or physical – remains. Nonetheless, there are some clues regarding the causes of China's deforestation. In the early fifteenth century China was a leading maritime power, assembling a fleet of over 300 gargantuan ships (manned by some 27,000 sailors) to explore the expanses of the Indian Ocean basin. The construction and maintenance of what was at the time the world's largest ocean-going fleet would have eaten deeply into its forest resources. The sheer size of China's population, which increased from 85 to 350 million between 1400 and 1800, meant that both its temperate and tropical forests were felled rapidly for building materials, firewood and charcoal and for agricultural expansion. The primary driver of Chinese deforestation was agricultural clearing, much of which took place in the seventeenth and eighteenth centuries during the ongoing process of internal colonisation. For example, as the Lignan region of south China (an area roughly the size of France) 'filled up' with Han immigrants from the north, large numbers of new settlers – spurred on by the commercialisation of the imperial state economy and the spread of New World crops, such as maize and sweet potatoes, that had enabled the cultivation of upland areas – soon converted its forests into farmland. Rice production for subsistence predominated, but many Chinese peasant farmers also specialised in cash crops such as tea and sugar which were traded on international markets. By 1800, agricultural expansion and relentless population growth had eliminated most of China's original forest cover (only remnants were left in the north-east and south-west corners of the country).

Having discussed the most important direct drivers of deforestation, we must now briefly examine the indirect drivers that underpinned the process. Population growth was a highly significant factor not only in China's deforestation but also in the destruction of forests worldwide. Table 3.2 below shows the remarkable rise in human population from 10,000 years ago to the present.

From the fifteenth century, accelerating population growth undoubtedly increased rates of deforestation. As societies grew, they cleared more forestland for food production and consumed more wood. But the relationship between rising population and environmental degradation is by no means a simple one. As the earth's population almost quadrupled in the twentieth century, forests in western Europe, North America and Japan actually began to recover, with regrowth now slightly exceeding timber extraction. Between 1910 and 1979, for example, in the USA over 200,000 square kilometres of its eastern forests either regenerated or were replanted. Slow population growth in advanced 'Western' countries, better forest management, rising recreational value, the abandonment of marginal farmland, the shift to fossil fuels and the substitution of metal and plastic for wood as a construction material all eased the pressure on their temperate forests. Above all else, the

Table 3.2 Estimated global population (in millions)

Date	*Population*
10,000 BCE	4
1 CE	200
1000 CE	270
1400 CE	370
1600 CE	550
1800 CE	920
1900 CE	1,625
2000 CE	6,000

Source: adapted from Joel E. Cohen, *How Many People Can the Earth Support?* (New York: W. W. Norton, 1995).

forests of the developed world – particularly in Europe and Japan – have rebounded because more and more 'cheap' food and timber have been imported from the developing regions of the tropical world. As around 80 per cent of a projected world population of 8 billion in 2020 will dwell in tropical countries, the onslaught on their forests looks set to continue.

Innovations in science, technology and transportation have had a profound effect on rates of deforestation, especially since the Industrial Revolution. Until the second half of the nineteenth century, felling trees was a back-breaking and time-consuming task. Pre-industrial clearing was heavily reliant upon human muscle and animal power, as trees were cut with axes and hauled out of the forest by horses or oxen. The improvement and extension of a versatile transport infrastructure (navigable rivers, canals, railways, roads and cable harvesting equipment) speeded the flow of logs to and from commercial sawmills. Improved transportation systems reduced haulage costs, important for the movement of low-value goods in large quantities, and their rapid extension meant that few of the world's forests were too remote to be harvested. The application of steam power in sawmills, which became commonplace from the 1870s, also enormously increased timber production. In the USA alone the amount of timber cut more than doubled from 20 billion board-feet to almost 46 billion board-feet between 1880 and 1906. But the ongoing 'industrialisation of the forest' did not extend to the actual felling of trees until after the Second World War, with the development of a practical chainsaw. First patented in 1858, it was not perfected until 1947. The adoption of the chainsaw allowed lumbermen to cut trees between 100 and 1,000 times faster than with axes. New technologies made the extraction and supply of timber both quicker and easier, and the development of new techniques to convert pulpwood into paper and packaging meant that all trees had a commercial value and could be exploited. Plant transfers, directed by scientists at botanical gardens throughout Europe and European colonies, also had a devastating impact on the world's forests. Economic botany – the organised intercontinental transfer of profitable plantation-grown species such as sugarcane, tea, coffee, cotton and rubber – produced 'green gold' for European empires, but at a high ecological cost (as discussed above).

Natural forces – climate, fire and disease – also induced dramatic changes in the world's forest cover. Some 10,000 years ago, as the climate shifted from glacial conditions to warmer temperatures at the end of the Ice Age, forests vastly extended their range. In Europe, for example, the boreal forest migrated northwards into Scandinavia and Russia, rapidly colonising tundra and steppe. As the world has continued to warm, however, partly as a result of anthropogenic activity, it is likely that the distribution of forests will alter again (with cool coniferous forests shifting even further northwards). Warmer temperatures also create the dry conditions that lead to fires, which can severely damage tropical, temperate and boreal forests. Fire is a natural component of many forest ecosystems, important in encouraging new growth and diversity. But although forests fires occur naturally, as a result of lightning strikes for example, since the capture of fire most have been started either accidentally or deliberately by humans. Uncontrolled burning can have catastrophic effects. In 1825, for instance, clearing fires begun by Canadian settlers escaped their control and consumed around a quarter of New Brunswick's forests. In Canada, wildfires still consume far more trees than timber companies (about 2 million hectares per annum), and two-thirds of the 9,000 or so conflagrations every year are caused by people. Outbreaks of disease also decimated forests in a short span of time and, as with 'virgin soil' epidemics among humans, they were often introduced accidentally from overseas into tree populations with no resistance. In 1904, for example, it was discovered that imported trees from Asia had brought chestnut blight to the USA. By 1940, some 3.5 billion American chestnuts had been lost to the fungus *Cryphonectria parasitica,* turning a once dominant forest tree into a threatened species.

Lastly, as we have seen, cultural values were a powerful driver in changing the world's forest cover. And when European settlers arrived in their new lands from the 1500s, many believed that it was their Christian duty to clear forests and convert them into fruitful farms as quickly as possible. The Judaeo-Christian tradition, which stressed humankind's right to dominate and subdue the natural world (*Genesis* 1:26–9), made transforming forest 'wilderness' into an earthly Eden a 'divinely ordained task'. In addition, fears that forests harboured 'wild people', criminals and other social outcasts prompted clearing to bring order to dangerous environments. Removing trees not only assisted the spread of settled agriculture, it was also a way of advancing effective colonial government by eradicating sites of disorder and resistance. Despite many writers suggesting that non-Western religious attitudes were more respectful towards the environment, as the experiences of China and Japan have shown, Buddhist, Confucian and Shinto 'reverence for nature' was not strongly reflected in their forest policies. Such religious values did not save forests, except for a few isolated stands of trees left untouched near shrines or other sacred places. Indeed, according to Conrad Totman, the construction of monumental Buddhist temples and Shinto shrines heavily depleted Japan's forests. Forest conservation on a global scale had its origins in European management regimes, complicating the notion that Western attitudes towards nature were fundamentally more hostile than others.

Forest conservation

Recent scholarship has suggested that the roots of modern environmentalism can be found in European approaches to forestry (as they anticipated many of today's ideas about resource management and climate change). The dynamic forces unleashed by European trade expansion, imperialism and industrialism severely depleted the world's forest resources. However, concerns about the consequences of deforestation led to the emergence of a conservationist movement. From the seventeenth century, reports from colonial scientists on St Helena and Mauritius – remote islands used as provisioning stations for ships on the main trading routes to India and China – linked the disappearance of their forests with desiccation, soil erosion and water shortages. The deleterious effects of deforestation were highly visible on small insular environments, and the ongoing importance of St Helena and Mauritius as repair and replenishment points for the ships of European East India companies prompted early attempts to set up forest-protection programmes. The pioneering work of Richard Grove posits that scientific ideas about forest conservancy originated at the peripheries of empire in these 'tropical island Edens'. Most environmental historians, however, have looked instead to the metropolitan centre for the beginnings of 'sustainable' forestry.

By the early modern period, laws regulating forest use had become commonplace throughout Britain, France and Germany (especially in woodland reserved for royal hunting). That forest clearance was the cause for widespread concern in north-western Europe is evident in two landmark documents: John Evelyn's *Sylva; or, A Discourse of Forest Trees and the Propagation of Timber in His Majesty's Dominion* (1664) and Jean-Baptiste Colbert's 'Ordonnance des Eaux et Forêts' (1669). Evelyn's *Sylva,* produced at the request of the British Navy, has been described as 'the first modern forest policy report'. From the early 1660s, Colbert, administrator of French forests under Louis XIV, systematically assessed the state of the country's timber (and water) supplies and how to better meet its needs, which resulted in the promulgation of the great Ordinance. Both Evelyn and Colbert recommended replacing uneven local regulation with a uniform national approach to improve forest management, as well as widespread replanting to ensure the long-term provision of essential wood supplies – particularly oaks for building warships. In Britain, a powerful landed elite hindered the development of an effective national forest policy. But in France, with its more centralised bureaucracy, the Ordonnance des Eaux et Forêts (comprising some 500 articles) became the 'bible' of forestry until after the 1789 Revolution. By the eighteenth century, however, German princedoms led the field where the 'scientific' management of forests was concerned.

German methods and approaches to forestry grew out of cameralism, a public policy doctrine that aimed to stimulate and modernise trade, industry and agriculture. Worries that wood shortages threatened Germany's economic health had encouraged the development of scientific forestry, based on the following agenda:

- detailed survey and inventory techniques;
- calculation of rates of tree growth;

- sustained yield management;
- replanting to regenerate depleted forests;
- strict regulation of forest use;
- specialised education and training for forest officials.

From the latter half of the eighteenth century, professional foresters began to map out the full extent of Germany's forests, provide data on different types of tree species, measure their various growth rates and determine how much timber could be harvested not just in the immediate future but over the course of a century or more. Scientific forestry, while seeking to address the environmental problems caused by overexploitation, was mainly concerned with sustainable resource management to meet different economic and material needs. The goal was to obtain a sustained yield from permanent forests using a range of silvicultural techniques, such as coppicing (regeneration through shoots from cut stumps) and shelterwoods (a system where new growth from seed or replanting is protected by stands of mature trees). As Michael Williams has noted, the standing forest was seen as capital and its yield as interest. It was a highly utilitarian approach to conservation, and the quantitative methods of German forestry, which provided state officials with an easily intelligible forest 'balance sheet', were to underpin management regimes across much of the globe by the early twentieth century.

In 1763, the first specialist school to teach scientific forestry was established in the Hartz mountain region of Germany. The most influential educational and research institutions in *Forsteinrichtung* (forest management) were founded in Tharandt, Saxony (1811) and Eberswalde, Brandenburg-Prussia (1830). Trained experts transformed damaged and disorderly forests into well-managed concerns, producing a reliable sustained yield that established Germany's reputation as world leader in 'economic forestry'. Other nations were keen to improve and protect their forests – and to make them profitable – and a steady stream of students from throughout Europe and its colonies were sent for training in German forest schools. The first director of the French National Forestry School, founded in Nancy in 1824, was the German-trained J. Bernard Lorentz, who actively promoted the scientific forestry agenda. During the nineteenth century, forestry schools were also established along German lines in Austria, Britain, Russia and Spain. Moreover, German experts were hired to help set up forest conservation and administrative systems around the world, such as Dietrich Brandis in British India (1864) and Bernhard Fernow in the USA (1886). By the late 1920s, there were fifty separate forest departments that served the British Empire, managing about 1.2 million square miles of revenue-producing forests from India and South Africa, to Australia and Canada. In 1906, the area of reserved forests managed by the Forest Service in the USA was a massive 60.7 million hectares. Forestry in Europe, the USA and much of the wider world shared common roots. Indeed, in 1916, China introduced a forest-management and regeneration policy modelled on Western scientific techniques.

Forest conservation, however, was not exclusively concerned with sustained yields and enhancing revenues. Long-term thinking about forest sustainability also incorporated non-timber values. From the seventeenth century, British, Dutch and French colonial scientists had pushed forest conservation forward on environmental grounds. But it was the American George Perkins Marsh's book *Man and Nature; or, Physical Geography as Modified by Human Action* (1864) that first alerted a global audience to the dangers of misusing forest and other natural resources. 'The earth', he warned, 'is fast becoming an unfit home for its noblest inhabitant.' His influential study, designed to show that humans had shaped the earth more than it shaped them, stressed the value of tree cover in protecting watersheds, preventing floods and soil erosion, providing habitat for wildlife and regulating regional climatic conditions. Marsh had faith that science could restore degraded nature, and he argued persuasively that what are today called 'ecological services' should be better protected by policy-makers. Furthermore, the naturalist and transcendentalist writer John Muir, co-founder of the conservationist Sierra Club in 1892, advocated preserving the beauty of forests in their 'natural state' for future generations. Forests not only had value in themselves, he argued, but were important recreational resources for those who sought solitude or adventure in 'wilderness' areas. But for most contemporaries, forest conservation meant the sustainable production of wood 'crops' to meet national needs. In 1906, Fernow had declared: 'Forests grow to be used. Beware of sentimentalists who would try to make you believe differently.'

By the early twentieth century, forest 'balance sheets' indicated that the consumption of wood in the USA and many of the imperial nations of western Europe was likely to outstrip sustainable production, causing concerns over a predicted 'timber famine'. Deforestation was increasingly viewed as a problem of worldwide importance. Reflecting these concerns, the US Forest Service undertook a global stocktaking of wood supplies, *The Forest Resources of the World* (1910), to evaluate how much standing timber was still available for use. In Britain, which had suffered severe wood shortages throughout the First World War as markets were disrupted, forest policy aimed at turning the empire into a self-sufficient timber-trading unit. In 1920, the first British Empire Forestry Conference was held in London. Others followed in Canada (1922), Australia and New Zealand (1928), South Africa (1935), and in London again (1947). Before the First World War, forest departments in different parts of the vast British Empire had worked in isolation from each other. These conferences brought them together, enabling scientific experts to share information and to coordinate forest-management strategies on an international scale. A number of important institutions were established on the recommendation of the British Empire

Forestry Conferences, including the Imperial Forestry Institute, the Commonwealth Forestry Bureau, the Forest Products Research Laboratory, the Tropical Products Institute, the Timber Development Association and the British Paper and Board Industry Research Association, some of which remain influential today under different names. The need for transnational cooperation in forest research had been recognised as early as 1892, when the International Union of Forest Experiment Organisations was established in Eberswalde, Germany, in response to major floods in central Europe thought to have been caused by deforestation. Delegates from just five nations – Austria, France, Germany, Italy and Switzerland – attended its first meeting. Today, renamed the International Union of Forest Research Organisations, it unites a network of 15,000 scientists from some 700 institutions in over 110 countries who are working to provide the knowledge needed to better protect trees and forests. But it is fair to say that over the past two centuries the impact of scientific forestry has been mixed in terms of its success.

The developing global forestry community pressed governments – more interested in timber production than environmental issues – to protect forest cover because it helped to safeguard the whole 'household of nature' (soil, water, wildlife and climate). Working plans employed by professional foresters set limits to forest use, considered environmental, economic and cultural needs, replenished standing stock by replanting and sought to prevent destructive fires and disease outbreaks. The new concept of sustainable yield meant that the old 'cut out and get out' approach to timber extraction, to borrow Williams' pithy phrase, faded in the USA, Europe and their colonies in favour of maintaining forests in perpetuity. Forests were increasingly viewed as 'tree farms', rather than timber 'mines' to be plundered. Tellingly, when the US Forest Service was established in 1905, it was as an agency of the Department of Agriculture. Monocropping was (and still is) the hallmark of modern agriculture, and as the primary aim of forest conservation was a sustained yield there was little incentive to maintain species diversity. To maximise productivity, cut-over land was often replanted with single-species timber crops in plantation style; neat rows of even-aged trees that could be efficiently harvested in set rotations, reducing the formerly rich composition of old growth forests to a sterile uniformity. Moreover, the widespread utilisation of fast-growing (often exotic) species, such as pines, spruces and eucalypts, improved output but could degrade water and soil resources as well as diminish food and fodder supplies for forest-dependent communities. By introducing monocultures, scientific resource management actually led to a decline in the quality of much forestland. And economically efficient annual clear-cutting left soils dangerously exposed and damaged the recreational value of forests. In practice, forest planning often struggled to reconcile different aims and interests. As ecological science firmly established itself during the 1920s and 1930s, more 'natural' approaches to forest management began to emerge. For example, the German *Dauerwald* (continuous forest) model, developed by Alfred Möller, called for uneven-aged and mixed-species planting to create both sustainable and biologically diverse forests. It also rejected clear-cutting methods of timber harvesting. But it was not until 1992, when the United Nations Conference on Environment and Development at Rio de Janeiro redefined the terms of sustainable forest management to include greater biodiversity, the promotion of native tree species, selective harvesting techniques and continuous tree cover that the *Dauerwald* model started to become more influential worldwide.

Around the world, the implementation of scientific forestry also involved restricting access to forests to ensure a sustained yield of timber. As farmers and lumbermen were perceived to exploit forestland wastefully for short-term gains, the state intervened to protect forests over the long-term. In Europe, strictly enforced forest-protection laws had been an integral part of the scientific resource-management agenda from the outset. This could be the source of conflict. In Germany and France, for example, peasant protests followed the passage of legislation that constrained traditional communal rights of forage, gleaning and pasturage. During the nineteenth century, colonial states enacted similar forest laws that curtailed access, limited usage and created forest reserves in their lands overseas including: the British Cape Colony in South Africa (1859), the Dutch colony of Java (1860), the French colony of Cochinchina (1862) and German East Africa (1893). Eradicating shifting cultivation, also known as swidden agriculture or slash and burn, was a priority for European-trained forest officers, especially in sub-Saharan Africa and south Asia. Despite being successfully practised by indigenous peoples for centuries, to European eyes shifting cultivation – an agricultural system where a small area of forest is cut, burned to make ash fertiliser, and crops raised for a few years before moving on – appeared backward, environmentally damaging, wasteful of valuable timber and politically disruptive as it encouraged 'nomadic habits'. Although some officers were aware that shifting cultivation could be an effective means of rejuvenating forests if sufficiently long rotations were employed, colonial states usually banned or restricted this 'primitive' form of agriculture, causing widespread discontent among rural populations. As had been the case in Europe, the forests of empire became sites of conflict as new regulations sought to curtail customary rights of use. Forest conservation, as Ravi Rajan points out, came to represent 'both injustice and disinheritance' as common resources were appropriated by the colonial state.

Legislation to protect forests was promulgated outside of Europe and its colonies too. In 1891, the Forest Reserve Act allowed the US Federal Government to create an extensive system of national forest reserves from land in the public domain. Concerned about rising rates of deforestation in a 'rapacious laissez-faire economy', it was soon followed by the Forest Management Act of 1897, which defined the purpose of the new reserves as being to 'improve and protect the forests' and to 'secure favorable conditions of water flow and furnish a continuous supply of timber' for US citizens. In the USA, the policy of reservation became a contentious issue as homesteaders, logging companies, sportsmen and others expected unfettered use of the forest by right. It also led to a rift in the early American conservation movement, as Gifford Pinchot, the head of the US Forest Service, worked closely with lumbermen and the public towards the 'wise use' of natural resources, while preservationists such as John Muir believed that the nation's forests should be left untouched in their 'natural state'. However, it was not just Westerners who were rethinking their relationships with the natural world. Determined to avoid foreign contacts, Japan developed its own forest-conservation ethic from the late seventeenth century. Recognising that resource preservation was important in maintaining its independence, the Tokugawa shogunate imposed tight controls on forest use by the Japanese people, high and low alike. Monumental building projects, both secular and religious, were 'nearly abandoned' as a timber-rationing system was applied even to high-ranking families. But forest regulation deprived poor villagers in particular of traditional user rights to gather fuelwood and other subsistence goods, frequently leading to disputes over resource allocation. Plantation silviculture also became commonplace throughout the islands, such as even-aged stands of indigenous sugi conifers (*Cryptomeria japonica*), to sustain timber output over the long term. The introduction of effective legislation and tree-planting policies helped to restore Japan's forests and to produce the 'green archipelago' that we know today. State intervention played a key role in forest conservation and regeneration worldwide, although the implementation of regulated, sustained-yield management systems by governments (national and colonial) almost always caused social conflict and ecological simplification – as the case study below exemplifies.

Case study: forestry in British colonial India

British colonial rule from the middle of the eighteenth century marked a watershed in India's environmental history as the intensity of resource extraction increased and forests came under pressure from expanding agriculture (although the subcontinent was never to be transformed into a neo-Europe). During the Mughal era, royal and state monopolies over valuable commodities such as ebony, sandalwood and teak had led to some over-harvesting, but, overall, a 'rough equilibrium' existed between the needs of Indian society and the availability of forest resources. Many people made a living from the forest, and traditional common-property regimes, overseen by local elites and village councils, controlled rights of access and use. The British conquest of India had a major impact on its forests and local communities' relationships to them. Under the British Raj (imperial rule), forestland was cleared to create extensive plantations of tea, coffee, cotton, sugarcane and other 'cash crops' for sale on metropolitan and world markets. By 1900, for example, there were 764 tea plantations in Assam alone, producing 145 million pounds of tea every year for export (it had long since become Britain's unofficial national drink). The growth of plantation agriculture in India substantially reduced forest cover, particularly in the north-eastern and southern regions of the subcontinent. However, as India's population increased from around 120 million to 280 million between 1700 and 1900, the less obtrusive expansion of peasant agriculture caused ever-advancing deforestation. For example, in the Sunderbans – the Ganges-Brahmaputra coastal delta – peasant rice production led to the gradual decline of the largest mangrove forest in the world. During the twentieth century, livestock numbers began to grow at a faster rate than the human population, which caused the conversion of woodland to pasture. In addition, India's forests were also coming under pressure from escalating British demands for timber products.

From the 1790s, India became an important supplier of timber to the British, as wars with Napoleonic France disrupted imports of wood from the Baltic and New England. Hundreds of different types of trees were identified by botanists (or, to use a current term, 'bio-prospectors') surveying India's temperate and tropical forests, but only a select few species – most notably deodar (*Cedrus deodara*), sal (*Shorea robusta*), teak (*Tectona grandis*) and sandalwood (*Santalum album*) – were preferred for harvesting to meet the combined commercial and defence needs of the British Empire. Deodar cedars, native to the Himalayas, could reach a height of around 200 feet and were highly valued for providing timber for ship masts and durable beams for the construction industry. Sal trees, which also grew predominantly in the Himalayas, were fast-growing and could be used for all types of construction. Timber from the teak forests of south-central India and the Malabar coast was crucial for ship construction in both Indian and British dockyards. A useful indicator of the growing demands placed on India's forests by the imperial shipbuilding industry is the increase in tonnage of Britain's merchant marine, up from 1.28 million tons in 1778 to almost 5 million tons in 1860. High-quality sandalwood from Mysore in south India was also cut and exported to China, where it was made into incense for use in religious ceremonies. However, it was the inability of India's timber supplies to keep pace with the rapid expansion of its rail network after 1853 that precipitated the introduction of European-style forest management to the subcontinent.

Expanding the rail network, crucial for moving troops around and for transporting goods to markets, led to timber shortages and the creation in 1864 of the Indian Forest Service (IFS) with the German expert Dietrich Brandis as its first inspector general. Between 1860 and 1910, India's railway system experienced dramatic growth from 1,349 kilometres to 51,658 kilometres of track, making it the fourth largest in the world by the latter date. The railway was a voracious consumer of timber resources, and by the 1870s it required more than 1 million sleepers annually. Deodar, sal and teak were the only native species tough enough in their natural state to be used as railway sleepers, and even they had to be replaced every six to twelve years. Following 'scientific' methods, Brandis and his successors in the IFS sought to meet imperial demands for essential timber products by converting India's mixed forests into monocultures of deodar, sal, teak and a few other favoured species like the chir pine (*Pinus roxburghii*). Evergreen Himalayan oaks, for example, often the only source of fodder and grazing for livestock in the winter months, were simply too slow-growing to be successfully adapted for commercial forestry. The European-trained officers of the IFS transformed Himalayan mixed oak-conifer forests into single-species stands of quick-growing chir pine, which provided valuable timber and resin products. But chir pine monocultures were of no use for feeding livestock, which undermined local village economies. There were also experiments with fast-growing exotic plantation species, such as Australian eucalypts, that generated healthy short-term profits. But the ability of 'gum trees', as eucalypts were also known, to soak up water and to encourage great fires caused serious ecological problems in semi-arid regions over the long run. However, while the strategic and commercial priorities of scientific forestry in India have been criticised for their negative social and ecological impacts, not least of all on biodiversity, recent research has stressed that much of the area set aside for forest management was already suffering from overcutting, overgrazing, soil erosion and fire damage. Before IFS control was established, local elites and timber merchants in late-eighteenth-century and early nineteenth-century India had begun to exploit its forest commons with little regard for the future. In contrast, Gregory Barton has argued that the IFS had a genuine commitment to environmental conservation and that, despite its economic *raison d'être,* it often resisted the demands of the colonial government for quick profits. For example, the working plans for the depleted Mudamalai, Kumbarakolli and Benne forests in the Nilgir district of Madras province, drawn up in 1909, show that the IFS expected to manage them at a loss for fully 100 years while they recovered.

Soon after the IFS was established, the colonial government passed two Indian Forest Acts, in 1865 and 1878, which created state-owned forests from what had been common property. The Acts divided India's forests into three main categories, reserved, protected and village (although the latter category was rarely used), and defined who had access to their resources. The State exercised close control over reserved forests, and public rights of access were either very limited or completely extinguished. In protected forests, the existing user 'rights' of rural people were recognised and registered but downgraded to the status of 'privileges' (which might be revoked when forests were closed for harvesting and replanting). Moreover, it was up to Indian peasants to prove that they had held customary forest rights in the first place, and many – illiterate and unfamiliar with Western concepts of private property – failed to legally register their claims. The settlement of local 'rights' under the Forest Acts clarified who had access to specific land areas for a range of purposes, from cutting timber and fuelwood to grazing livestock and gathering minor forest products such as fruits and honey, enabling the IFS to manage India's forests on a sustained-yield basis for 'the benefit of all'. But since the expert view of IFS officials was that 'reserved' status offered the best protection, by 1899, some 81,400 square miles of a total of 90,200 square miles of state-owned forests had been converted into this category. This vast forest system covered at least 20 per cent of the land area of British India, and stringent regulations regarding its use affected the lives of most rural dwellers. India's 'multiple-use' approach to forest administration was considered a model of good practice, and similar systems were established throughout the British Empire and the wider world. By 1900, forest departments had been set up with the assistance of the IFS in Australia, the Cape Colony, Ceylon, Cyprus, the Gold Coast, Kenya, Malaya, Mauritius, New Zealand, Nigeria, Sierra Leone, the Straits Settlements and Uganda. Significantly, Dietrich Brandis trained several officers for the US Forest Department, including Gifford Pinchot. Expertise drawn from India played a major part in the development of a global forestry community.

The implementation of new forest laws, however, caused widespread discontent throughout the Indian subcontinent. While many peasants still had access to forest resources for domestic consumption, they no longer enjoyed the right to sell firewood, fodder or minor forest products to supplement their meagre incomes. Hunter-gatherer communities, such as the Chenchus of Hyderabad, found that their traditional subsistence activities were now forbidden in state forest reserves. There were also attempts to restrict shifting cultivation, which IFS officers considered wasteful of valuable timber resources. In the name of conservation, thousands of IFS guards enforced laws that criminalised most customary uses of the forest. Several decades after the passage of the 1878 Act, recorded 'forest crimes' were averaging 30,000 per annum in the province of Madras alone, demonstrating that the peasantry did not surrender their rights easily. Conflicts over access to forest resources also reflect the great unpopularity of state intervention among agrarian society. Resistance to the State's forest policy took many forms, from peacefully petitioning the local administration to the setting of fires and outright rebellion. Unsurprisingly, arson attacks targeted deodar, sal, teak and pine forests: species that had high commercial value to the State but which were of less use to rural populations. Forest grievances sparked many small risings or *fituris,* such as the Rampa rebellion of 1879–80 which took ten army companies to put down. The most common forms of resistance, however, were unspectacular and non-violent, including petitions, peaceful strikes and 'avoidance' migration to areas where the forest laws were less vigorously applied. The IFS sought to defuse hostility by setting aside some forestland for shifting cultivation. For example, Brandis developed the 'taungya' system of agro-silviculture, where shifting cultivators were allowed to clear forests to grow food provided that they replanted trees alongside, leaving behind a commercial timber crop when they moved on. The taungya system reduced social tension and became an important plantation technique in India (and several other Asian and African countries). Labour by shifting cultivators on behalf of the IFS, however, was generally unpaid.

The loss or restriction of customary user rights under the new laws forced many forest-reliant people off their lands to seek work as agricultural labourers on farms or plantations. The economies of whole villages suffered too when mixed forests were replaced by economically efficient monocultures. Indigenous knowledge about protecting and sustaining forest habitats, acquired over centuries, was disregarded by IFS experts. Timber production was the primary aim of 'multiple-use' forest management in India, and the needs of the colonial state were prioritised over those of local communities. The balance sheet for scientific forestry in India ultimately shows that the British Raj drew too heavily on its natural capital, especially during the First and Second World Wars when the abandonment of working plans led to large-scale deforestation. After India won independence in 1947, its rural dwellers were hopeful that traditional forest user rights would be restored. But Indian forest policy remained fundamentally unchanged. In 1952, the Government of India issued a policy statement which emphasised that forests would be managed to meet national, rather than local, resource needs. It reinforced the right of the State to exclusive control over commercial timber production, and, during the post-independence drive for industrial development, forests were viewed mainly as the source of raw materials and commodities to sell on world markets to generate much-needed revenue. In particular, demands from its new paper, pulp and plywood industries placed added pressure on India's forests. Conservancy measures, closely identified with the old colonial regime, were largely ignored as the State pressed for higher timber output from forestlands that were still recovering from wartime abuse.

The continued commercial orientation of Indian forestry provoked renewed conflict with village communities concerned to protect their livelihoods. The famous Chipko *andolan* of the 1970s and 1980s is perhaps the best-known forest protest movement. Led primarily by women, the peaceful Chipko, or tree-hugging, movement defended customary forest rights in the central Himalayas against the incursions of commercial timber operators. In comparison with earlier peasant protests, Chipko activism was distinctive because – attuned to the developing environmental consciousness of the times – it promoted the message of sustainable forest use (for fuelwood, timber, foodstuffs, fodder and the maintenance of 'ecological services'). It helped to inspire the creation of Joint Forest Management schemes in the late 1980s, which involved local communities in decision-making and gave them a more equitable share of forest resources. As a result of such community forestry initiatives, which have now spread across twenty-seven Indian states, significant strides have been made in promoting cooperation between poor villagers and forest departments to plant and protect trees. The long-term goal, emphasised in the 1988 National Forest Policy, is to cover one-third of the subcontinent with forestlands. National Parks, conservation areas and bioreserves have also been created to provide a higher degree of protection for ecologically important forest habitats. But today, only around 11 per cent of the Indian landscape is thickly forested, of which many thousands of square kilometres are single-species timber plantations. Even community forestry, with its roots in the Chipko movement, has tended towards specialised tree-planting of commercial species such as teak, pine and eucalyptus. Less than 1 per cent of the country is covered by primary (naturally regenerated) forest. Given that India is now the world's second most populous country with around 1.1 billion people and a booming economy, the sustainable management of its remaining forest resources represents a considerable challenge for the future.

Conclusion

Clearing forests to extract wood and to create agricultural land has dramatically reshaped the face of the earth. In 1500, deforestation was hardly a new phenomenon, but after this date the development of a global market economy, closely connected to burgeoning population growth and the expansion of European imperialism, saw a local and national concern become a worldwide problem. Innovations in transport and technology have greatly accelerated the process over the past two centuries, particularly the widespread adoption of gasoline-powered chainsaws, tractors and trucks after the Second World War. Although deforestation is a complex process, 'operating at various scales, in various places, and with a multiplicity of variables', its underlying causes are generally agreed upon. It is also possible to discern global patterns of change. Temperate rather than tropical forests experienced the highest rates of clearing until the twentieth century, when the rise of the environmental movement in the developed world, among other factors, saw this pattern reverse. However, the recent regeneration of forests in the temperate North, especially in Europe and Japan, would not have been possible without imports of 'cheap' timber and food supplies from the tropical South – where deforestation rates are still rising. Around 1 per cent of the world's forests are still lost every year, mainly in ecologically vulnerable tropical zones – in Brazil, Cambodia, Indonesia, Myanmar, Papua New Guinea and other countries – where poor soils make regeneration more difficult.

Recent scholarship has shown, however, that environmental thinking (in the sense of 'wise use' conservationism) on a global scale had its roots in empire forestry. Scientific forest management, based on the Germanic idea of sustained yield, was exported not only to European colonies but also to many other countries, including the USA and China. This global forestry community shared a common agenda, but its implementation varied from place to place according to local ecological conditions and the prevailing political climate. It is also important to note that similar administrative systems could evolve in places outside of European influence, such as Tokugawa Japan. From the mid-nineteenth century, growing anxieties about timber shortages and the environmental effects of deforestation saw governments across the world take control of forests which were traditionally some form of common property, curbing rights of access and use with strict legislation. The utilitarian ethos of state-backed forest departments, which by the early decades of the twentieth century were cooperating internationally, created landscapes that were characterised by monocultural stands of preferred commercial tree species (modified or secondary forests and plantations). The impact of these institutions on the livelihoods of local and indigenous communities, who were often treated as destructive 'enemies of the forests', provoked widespread protests and lasting resentment. The multi-use approach to forestry, pioneered in British India, which attempted to balance demands for timber products with environmental protection, sport and recreation, and traditional socio-economic and cultural uses, did not work out particularly well in practice. The aims of forest departments were often 'contradictory and inconsistent', and the results were social conflict and a decline in forest area and quality. Despite a genuine commitment on the part of many professional foresters to long-term conservation, rising demands from all quarters meant that natural capital as well as its sustainable yield was generally consumed at a faster pace than it could be replenished.

Today, the demand for fuelwood, minor forest products and new land to cultivate remains high among the world's poorest people. The industrial consumption of timber – despite the introduction of new materials such as concrete and plastics – is still growing, with both China and India now leading importers of wood products. The pressures on global forests, particularly those in the tropical South, show few signs of abating. Timber plantations in both the temperate and tropical worlds, though they lack biodiversity, are now thought to have a crucial role to play in protecting the world's remaining 'old-growth' forests because of their high productivity. Around 10 per cent of the world's forests have been protected in National Parks, conservation areas or bioreserves, with most being found on the American continent. In recent years, governments in many parts of the world have also begun to include indigenous and poor village communities in participatory conservation schemes, recognising that their cooperation and intimate knowledge of local environments can lead to more sustainable forest management. The Joint Forest Management strategy implemented in India since the 1980s has proved to be an influential model in this respect. Timber products are also being certified by independent, non-profit organisations such as the Forest Stewardship Council to show that they have been sourced from forests managed in a sustainable way, although internationally recognised 'Green' trading standards have yet to be agreed. At the Rio 'Earth Summit' in 1992, environmentalists from the developed world suggested the establishment of an international management system to combat deforestation. However, this idea was swiftly rejected by delegates from countries in the developing world, as it smacked of neo-imperialism. Perhaps the most promising new initiatives that offer hope for the future are those that finally place an economic value on the 'ecological services' that forests provide both for the planet and the benefit of us all. Early in 2008 two ground-breaking schemes were set up by British-led partnerships to pay governments in central Africa and Guyana to protect tropical rainforests as part of international efforts to tackle climate change. A similar deal has also been brokered between Norway and Costa Rica. Such initiatives, which seek to halt deforestation by making trees worth more as living resources than as lumber, may help to reduce global warming and to prevent the current forest crisis from deepening.

Further reading

David Arnold, *The Problem of Nature: Environment, Culture and European Expansion* (Oxford: Blackwell, 1996).

Greg Bankoff and Peter Boomgaard (eds), *A History of Natural Resources in Asia: The Wealth of Nature* (New York: Palgrave Macmillan, 2007).

Gregory A. Barton, *Empire Forestry and the Origins of Environmentalism* (Cambridge: Cambridge University Press, 2002).

William Beinart and Peter Coates, *Environment and History: The Taming of Nature in the USA and South Africa* (London: Routledge, 1995).

William Beinart and Lotte Hughes, *Environment and Empire* (Oxford: Oxford University Press, 2007).

Warren Dean, *With Broadax and Firebrand: The Destruction of the Brazilian Atlantic Forest* (Berkeley, Calif.: University of California Press, 1997).

Ken Drushka, *Canada's Forests: A History* (Montreal and Kingston: Forest History Society/McGill-Queens University Press, 2003).

Ken Floyd, *Forest Sustainability: The History, the Challenge, the Promise* (Durham: Forest History Society, 2002).

Madhav Gadgil and Ramachandra Guha, *This Fissured Land: An Ecological History of India* (Delhi: Oxford University Press, 1992).

Tom Griffiths and Libby Robin, *Ecology and Empire: Environmental History of Settler Societies* (Edinburgh: Keele University Press, 1997).

Richard Grove, *Green Imperialism: Colonial Expansion, Tropical Island Edens and the Origins of Environmentalism, 1600–1860* (Cambridge: Cambridge University Press, 1996).

Ramachandra Guha, *Environmentalism: A Global History* (Harlow: Longman, 2000).

——*The Unquiet Woods: Ecological Change and Peasant Resistance in the Himalaya,* expanded edn (Berkeley, Calif.: University of California Press, 2000).

Rashid Hassan, Robert Scholes and Neville Ash (eds), *Millennium Ecosystem Assessment, Volume 1: Ecosystems and Human Well-Being – Current State and Trends* (Washington, DC: Island Press, 2005).

Alf Hornberg, J. R. McNeill and Joan Martinez-Alier (eds), *Rethinking Environmental History: World-System History and Global Environmental Change* (Lanham, Md.: Altamira, 2007).

J. Donald Hughes, *An Environmental History of the World: Humankind's Changing Role in the Community of Life* (London: Routledge, 2001).

Shephard Krech, John McNeill and Carolyn Merchant (eds), *Encyclopedia of World Environmental History* (New York: Routledge, 2004).

John McNeill, *Something New under the Sun: An Environmental History of the Twentieth Century* (London: Allen Lane, 2000).

Robert B. Marks, *Tigers, Rice, Silk and Silt: Environment and Economy in Late Imperial South China* (Cambridge: Cambridge University Press, 1998).

Nancy Peluso, *Rich Forests, Poor People: Resource Control and Resistance in Java* (Berkeley, Calif.: University of California Press, 1992).

Ravi Rajan, *Modernising Nature: Forestry and Imperial Eco-Development 1800–1950* (Oxford: Oxford University Press, 2006).

Emil Salim and Ola Ullsten (eds), *Our Forests, Our Future: Report of the World Commission on Forests and Sustainable Development* (Cambridge: Cambridge University Press, 1999).

Conrad Totman, *The Green Archipelago: Forestry in Pre-Industrial Japan* (Athens, Ga.: Ohio University Press, 1998).

B. L. Turner, William C. Clark, Robert W. Kates, John F. Richards, Jessica T. Mathews and William B. Meyer (eds), *The Earth as Transformed by Human Action* (Cambridge: Cambridge University Press, 1990).

Michael Williams, *Americans and their Forests: A Historical Geography* (Cambridge: Cambridge University Press, 1989).

——*Deforesting the Earth: From Prehistory to Global Crisis* (Chicago, Ill.: University of Chicago Press, 2003).

Donald Worster (ed.), *The Ends of the Earth: Perspectives on Modern Environmental History* (Cambridge: Cambridge University Press, 1988).

Chapter 4

Soils and irrigation

Nothing has been more fundamental to the success of human societies than their relationships with the earth's soils. Unique to the earth, as far as is known, soils are a vital natural resource that holds the nutrients, moisture and microorganisms essential for global food production. Since the emergence of agriculture in the Fertile Crescent of the Middle East some 11,000 years ago, the dual challenge facing farmers everywhere has been how to sustain soil fertility and prevent soil erosion. Agricultural activity has the potential to enrich or deplete soils, and with careful land management crops flourish and livestock thrive from generation to generation. But poor farming practices can soon strip a region of a crucial resource that is non-renewable over human timescales. Soils are formed slowly over centuries or millennia, accumulating as a 'living skin' on the planet's surface that is rarely more than a metre or two thick (and often much less). Agriculture makes soils vulnerable to erosion and exhaustion at faster rates than the processes that created them. Estimates for erosion vary widely, but soils are currently being lost at somewhere between two and ten times natural rates (and often more rapidly in tropical environments). Across the globe, human action has seriously degraded about 2 billion hectares of agricultural land, an area the size of the USA and Canada combined.

Since the dawn of agriculture, human-induced soil erosion has surged three times: first, around 2000 BCE as agriculture spread from the river valleys of China, the Middle East and South Asia; second, with the worldwide expansion of European settler societies from the fifteenth century; and third, after the Second World War, with the advance of commercialised agriculture onto the marginal lands of the tropics and the widespread adoption of heavy farm machinery. This chapter is chiefly concerned with investigating the impacts of the second and third surges, which greatly accelerated soil erosion on a global scale. The environmental changes caused by the use of irrigation systems and chemical fertilisers in producing crops and combating soil exhaustion are discussed. This chapter will also examine the evolution, implementation and success of soil management and conservation strategies. We will begin, however, by briefly examining the key role played by the world's soils in sustaining life.

Soils and the environment

Some of the earliest surviving writings deal with agriculture and, unsurprisingly, they recognise that in a rural world the wealth and well-being of societies derived mainly from the soil. For example, the Roman statesman Marcus Porcius Cato offered businesslike advice to the empire's farmers on how to get the most out of their land in his book *De agri cultura* (c. 150 BCE). Cato understood that particular crops grew best in particular soils, such as olive trees on rocky ground, and he distinguished twenty-one different classes of soil on that basis. Written around a century earlier, the ancient Chinese text *Guan Zi* identified no fewer than ninety distinctive soil types. Knowledge of soils and their use acquired during antiquity endured down into the early modern period. In 1679, echoing the earliest works, the English scientist John Evelyn advised agriculturalists 'to be well read in the Alphabet of Earths' to sustain good harvests. But the first extensive scientific survey was not undertaken until 1883 by the geologist Vasily Dokuchaev, who mapped and classified the rich chernozem soils of continental Russia. Widely acknowledged as the 'founding father' of soil science, he developed the basic system to describe the various layers or horizons of soil profiles, from topsoil down to bedrock. Dokuchaev's ideas became influential when in 1927 a book by one of his students, Konstantin Glinka's *The Great Soil Groups of the World and their Development,* was translated into English. Since then, most national soil classification systems have applied Dokuchaev's approach, and expanded on it, using a broad set of criteria based on colour, texture, structure and other physical and chemical properties when identifying and mapping soils. In the USA, for instance, the National Cooperative Soil Survey, founded in 1899, has now classified and mapped over 20,000 different kinds of soils.

Soils, then, are highly diverse – a global patchwork of contrasting types – and climate, topography, native vegetation and the underlying geology of a region all strongly influence their formation. Granite rocks, for instance, weather into sandy soils over time, while basalt gradually forms fine-textured clay-rich soils. The world's most fertile soils – light, wind-deposited loess varieties – are mainly to be found beneath the plains and prairies of North America, the pampas of South America and the steppes of Russia and Ukraine: held in place primarily by the root systems of perennial grasses. The Loess Plateau of north-western China, once densely forested, also provides good agricultural land. But globally only around 11 per cent of soils can be farmed without first being 'improved' in some way, such as through the development of irrigation systems in arid and semi-arid regions. Most of the planet's soils are less productive than loess and more difficult to farm, particularly those of tropical rainforests in sub-Saharan Africa, south and south-east Asia, and Central and South America which tend to be both shallow and nutrient-poor. The first Europeans to explore these forests were astounded by the lush vegetation they encountered and erroneously assumed that once the land was cleared their soils would be highly productive. But heavy rainfall, high temperatures and the severe weathering of parent rocks rapidly leaches out most of the nutrients from tropical soils. Fertility is largely retained and recycled by living vegetation and a thin layer of decaying matter (leaf litter) on the forest floor, and once rainforests are cleared the productive capacity of their soils is greatly reduced.

Before 1500, most cultivated land was to be found in the Old World (Europe, Asia and parts of Africa). However, over the past 500 years the massive expansion of commercialised agriculture has brought almost all of the world's suitable land into production. To meet burgeoning demand for food, as global population growth accelerated rapidly, marginal lands (tough to farm, with low long-term yields) have increasingly been converted to crops and pasture – especially after 1945 in the tropics. Today, around 35 per cent of the earth's terrestrial surface is farmed (11 per cent cropland; 24 per cent pastureland). The clearing of land for agriculture has been the primary means by which human societies have reshaped their environments over time.

Most people are aware of the importance of healthy soils in cultivating cereals, fruits and vegetables and in growing fodder crops and grasses to raise livestock, which have long been the staples of global food production. Soil supplies more than 95 per cent of our food. Besides food, the world's cultivated land also produces fibre crops (cotton, flax and jute), timber, oilseeds, dyestuffs and numerous other raw materials required by industry. Certain types of clays have been used for centuries to make bricks, tiles, pottery, porcelain and drainage pipes. Soils, however, have always served many other vital functions, and provided a range of 'ecological services', that are not so widely appreciated, including:

- **Connecting global systems:** Soils (the pedosphere) are in contact with the gases of the atmosphere, the groundwater of the hydrosphere and the minerals of the lithosphere, linking them together 'in one body' and supporting every form of life on earth (the biosphere).
- **Below-ground biodiversity:** Soils are home to myriad living organisms and microorganisms, from earthworms to bacteria. Just half a hectare of soil can contain around 125 million small invertebrates and countless bacteria, algae and fungi. Species diversity below ground greatly exceeds that above ground, which makes soils important repositories of biodiversity in their own right.
- **Nutrient cycling:** Essential nutrients for healthy plant growth – particularly nitrogen, phosphorus, potassium, calcium, sulphur and magnesium – are obtained from the soil. The nutrients available depend on the weathering of the soil's parent rocks, the actions of bacteria that 'fix' nitrogen to plant roots and other soil organisms that break down decomposing

organic matter into humus. Soil organisms play a crucial role in the cycling of nutrients and in creating fertile, humus-rich topsoil.

- **Water supply:** Keeping plants nourished requires a regular and ample supply of fresh water, either from rainfall or irrigation. Soils, especially those that are humus-rich, have the capacity to absorb and hold large amounts of water – serving as a kind of reservoir for use by plants. Their ability to retain water also reduces the risk of serious flood episodes. In addition, soils function as 'natural water treatment works', filtering out many dangerous pollutants before they can contaminate groundwater – helping to protect supplies of drinking water.
- **Carbon storage:** soils sequester more carbon than that found in the atmosphere and above-ground biomass combined, helping to combat climate change.

Biologically and organically rich soils provide multiple functions and 'ecological services' that benefit human communities – underpinning our very existence. Yet the study of soil–society relations, as John McNeill and Verena Winiwarter point out, is 'perhaps the most neglected subject within environmental history'. However, as contemporary concerns about rising rates of soil erosion, exhaustion and salinisation have grown, historical interest in human interactions with the land has begun to increase, focusing in particular on the role of agriculture in causing these environmental problems.

Losing ground: soil erosion, exhaustion and salinisation

Initially, it is important to note that most types of soil degradation were – and are – natural processes and that in some cases they could prove highly advantageous to human societies. The two main natural 'agents of erosion' were heavy rainfall and strong winds, both of which removed and redistributed large quantities of topsoil (but others such as glaciations, droughts, avalanches and even burrowing animals also moved earth around). Water erosion, for example, transported around a millimetre of fresh, nutrient-rich silt from the Ethiopian highlands to the Egyptian Nile Valley every year – this regular 'gift of the Nile' replenishing soils and sustaining agriculture on its floodplain for over five millennia. Elsewhere, the silt-laden floodwaters of the Tigris and Euphrates rivers in Mesopotamia, the Indus and Ganges rivers in the Indian subcontinent, and the Yangtze and Hwang-ho rivers in China were similarly vital to early farming activities, if less reliable in delivering their bounty. Wind erosion also carried fine soil particles over considerable distances with, for instance, the deep loess deposits of northwestern China, its prime agricultural region, probably having blown there from Mongolia. However, the expansion of settled agriculture resulted in a sharp acceleration of 'normal' water and wind erosion rates, as well as other forms of soil degradation.

Although it is not always easy to distinguish between natural and human-induced losses, agricultural practices have been closely linked to soil degradation. Clearing land for crops removes year-round vegetative cover, leaving bare earth exposed to the elements for months at a time, which results in erosion rates many times faster than the weathering processes that created soils. Under natural vegetation – grasses or trees – soils are better protected from the erosive action of water and wind, as root systems, leaf litter and closed forest canopies help to keep them in place. Topography is important too, as soils can be lost very quickly from steep hillsides – making upland farms more vulnerable to erosion problems than those situated in lowland areas. Soil exhaustion can usefully be considered as another type of erosion process, as continued cropping in the same location greatly diminishes the nutrients essential for healthy plant growth. Ploughing the land also reduces the numbers of soil-dwelling organisms that enhance its fertility. Without good drainage, the adoption of irrigation technologies – especially in arid and semi-arid regions – can also have adverse ecological consequences, causing waterlogging and salinisation (rising groundwater brings naturally occurring salts, harmful to most plants, to the surface). Poor livestock management on lands that are converted into pasture can lead to erosion problems, as overstocking of cattle, sheep, goats and other domesticated animals compacts soils, destroying their structure, and overgrazing removes the vegetation that holds the earth together. The malign effects of human-induced 'accelerated erosion' can be traced back to at least 2000 BCE, when cultivation-based economies emerged from the river valleys of China, the Middle East and South Asia.

Riverine agriculture depended on large-scale and complex irrigation systems, overseen by powerful ruling elites, to control, store and direct seasonal floodwaters and obtain more food from the land. It seems that the 'hydraulic civilisations' of Mesopotamia, India and China were 'stable and successful' for centuries. But as population growth began to outstrip the productive capacity of fertile floodplains around the second millennium BCE, sedentary agriculture spread from riverine sites onto the more vulnerable soils of former forestlands. Rich soils on alluvial floodplains formed relatively quickly, from the silt, clay and sand transported by great rivers, and were usually subject to low rates of erosion. In contrast, accelerated soil erosion soon became a problem in newly settled areas less suited to intensive farming practices. Land degradation on north-western China's Loess Plateau, as forests were cleared and grasslands opened up to rainfed agriculture, exemplifies the first surge of soil erosion in world history. The deep soils of the Loess Plateau, being both light and loose, were easy to cultivate and made fertile farmland. But as fields of millet replaced oak-dominated broadleaf forests and grasslands, rain and wind rapidly stripped this high, flat plain of soils laid down over millions of years. Without permanent vegetation to protect against erosion, heavy summer rains in particular removed millions of tonnes of fragile loess every year, with the muddy runoff giving the Hwang-ho (Yellow River) its name. Recently described as 'one of the world's most easily eroded landscapes', more than 70 per cent of the Chinese Loess Plateau is now scarred by deeply incised gullies and ravines.

Despite its initial stability and successes, over time irrigation agriculture in the Tigris-Euphrates, Indus and Hwang-ho river valleys also experienced major environmental problems. Accelerated soil erosion on China's upland Loess Plateau increased the incidence of serious flooding on the crowded Hwang-ho valley floor. Because of its enormous sediment load, the bed of the Hwang-ho constantly rose and shifted, which caused the river to burst its banks unpredictably – resulting in devastating inundations, destruction of crops and huge loss of life from both drowning and famine. In the past 3,000 years the Hwang-ho has changed its course twenty-six times and flooded more than 1,500 times, killing millions of people and earning the sobriquet of 'China's sorrow'. Droughts and low river flows were also problems for early irrigation networks, as weak floods clogged waterways with silt and failed to saturate and replenish the soil – meaning poor harvests and food shortages. But an equally potent threat came from growing soil salinity. Scholars such as J. Donald Hughes, David R. Montgomery and Clive Ponting have argued convincingly that soil salinisation, caused by inadequate drainage and overuse of the land, contributed to the decline of hydraulic civilisations in the southern Mesopotamian region of Sumer and the Indus valley of South Asia. Both the Sumerian (c. 3000–1800 BCE) and Harappan (c. 2600–1700 BCE) civilisations declined in relatively quick succession, partly because their salt-damaged soils were no longer able to support intensive agriculture. The first literate society, Sumerian writings (cuneiform script on clay tablets) recorded how their lands turned white, encrusted with salt, and how fields had to be abandoned because crops would no longer grow in saline soils. The failure of ancient civilisations to effectively adapt their agricultural systems to the environment meant that they were unsustainable over the long term. By examining cases in the distant past where complex, irrigation-based societies have 'ended up destroying what they created' (others include the Anasazi and Mayan civilisations of the Americas), authors such as Hughes, Montgomery and Ponting aim to raise awareness that our future well-being depends on sound stewardship of the soil.

The socio-environmental impacts of the first surge of soil erosion were serious and long-lasting but essentially regionalised. The second surge, set in motion by the expansion of European settler societies from the fifteenth century, was more profound and global in scale. Linking the first and second surges together – and the post-1945 surge too – is the notion that when people migrated to new lands they became 'environmental pioneers', encountering soils that were unfamiliar and often more easily eroded than those they had been used to working with. Importantly, northern European farming practices had evolved in a temperate climate, typically on gently undulating terrain with heavy soils that were not so vulnerable to erosion by wind and rain. But when European settlers applied their traditional farming techniques in the Americas, southern Africa, South Asia and Australasia – places with lighter soils, more intense rainfall regimes and steeper slopes – rates of soil loss could accelerate dramatically.

Erosion problems emerged soon after the Iberian discovery and conquest of the New World. The introduction of Merino sheep by Spanish settlers in Mexico's cool Valle del Mezquital during the sixteenth century, for example, triggered a complex set of actions and events that caused severe soil degradation in this formerly productive agricultural region. The indigenous Otomí people, their population decimated by recurring epidemics of Old World diseases, were quickly displaced from their lands by Spanish pastoralists and their domestic animals, most notably vast flocks of sheep. From the 1560s, prices for meat and wool to supply local, intercolonial and Spanish markets had begun to rise. Hills were deforested by these Spanish 'environmental pioneers', and native croplands converted to grasses, as large-scale sheep grazing took precedence over classic Mesoamerican agricultural production – the triad of maize, beans and squash. When grown together, this crop combination helps to both protect soils and sustain their fertility (tall maize stalks provide support for growing beans; bacteria living symbiotically in root nodules of legume crops like beans fix nitrogen into the soil; and large-leaved squashes provide ground cover and retain moisture). Despite its temperate climate, the Spaniards – who did not know the area like the Otomí – failed to turn the Valle del Mezquital into the archetypal verdant neo-European landscape. A 'plague' of alien Merino sheep (around 4.4 million animals by the 1580s) rapidly stripped the land of most of its vegetation, while their hooves trampled the bare earth, exposing fragile soils on high steep-sided hills to the erosive action of the wind and heavy summer rainfall. The desire for short-term gain had led pioneer colonists to ignore not only indigenous land-use systems but also their own customary rules controlling overgrazing. According to Elinor Melville, the Valle del Mezquital was actually closer to the European ideal of a flourishing agricultural landscape before the Spanish conquest. By 1600, the unintended outcome of applying Spanish stock-raising techniques in a new and unfamiliar environment had been to transform once rich and fertile farmland into one of the most gullied and barren regions of Mexico.

If the Iberians had led the way, by the late sixteenth century other European nations were establishing their own empire-based systems of agriculture, in pursuit of profit and greater food security. As European empires expanded around the world, forestland (see Chapter 3) and locally adapted indigenous agricultural systems that produced a diversity of subsistence crops were increasingly replaced by commercial monocrops for export, such as cacao, coffee, cotton, rice, sugar, tea, tobacco and wheat. Under colonial rule, the demands of European consumer markets dramatically reshaped environments, and little attention was paid initially to the ecological costs of development. The first European settlers of what Alfred Crosby called the 'neo-Europes' – Argentina, Australia, southern Brazil, Uruguay, New Zealand and much of North America – and resident plantation owners in the tropics imagined that their soils were inexhaustible, an unending source of prosperity and plenty. They also mistakenly believed that their traditional farming methods would be suited to the new conditions. But as animal-drawn ploughs displaced native digging sticks and hoes in field work, fertile topsoils became more vulnerable to erosion – especially on sloping ground. In addition, repeated plantings of the same crop fast depleted nutrient reserves. Indeed, many environmental historians have spoken of the 'mining' of soil fertility within colonial empires, as producing crops for sale on world markets removed valuable nutrients from local agro-ecosystems without adequately replacing them. The easy availability of fresh land meant that when agricultural productivity dropped, frontier farmers often moved on rather than spend time and money on improving and restoring degraded soils. It was labour, rather than land, that was in short supply at the frontiers of empire.

Over the past half millennium, as Richard Tucker has noted, plantation agriculture has been the 'most severe exploiter and transformer of both social and ecological systems'. More than 9 million African slaves were brought across the Atlantic to labour on commercial plantations in the New World. The human costs of coercive colonial agricultural systems have been recounted many times, but not the environmental consequences. The rise of the West's 'soft drug culture' saw sugar, tea, coffee, chocolate and tobacco become important commodities in world trade from the sixteenth century onwards, and forests and grasslands were extensively cleared to make way for export-oriented plantations. Iberian colonisers developed this intensive method of production during the second half of the fifteenth century, establishing small-scale sugar plantations in Madeira and the Canary Islands. In the 1570s, the first large-scale plantation was established by the Portuguese in north-eastern Brazil, with exports to Europe soon reaching 10,000 tons of sugar per annum. Within a few decades, the profitable Portuguese model of sugar monoculture was taken up by Dutch, British and French entrepreneurs, who transferred it to the mountainous islands of the Caribbean. Investment in irrigation schemes and the development of transport infrastructures, particularly the port facilities that linked New World plantations to the global economy, saw sugar exports to Europe rise to over 330,000 tons per annum by the early nineteenth century. Sugar grown on the 'ghost acres' of South American and Caribbean colonies provided a stimulating caloric supplement to dull European diets, at a time when malnutrition was still widespread. But intensive plantation agriculture was to have a devastating impact on New World environments.

Raising sugar year after year quickly robbed soils of their nutrients. In north-eastern Brazil, canefields were abandoned by Portuguese planters as 'exhausted' after just fifteen years on average, and more agricultural land had to be opened up at the expense of the Atlantic rainforest. Barbados, a small Caribbean island of only 440 square kilometres, offers perhaps the most thoroughgoing example of the environmental changes caused by the emerging plantation system. Indigenous *conuco* agriculture, a system of cultivating root crops in large earth mounds to slow erosion, was swept aside by British merchant-adventurers as Barbados became the primary site for their sugar operations. The wholesale conversion of the island to sugar production by British planters from the 1640s stripped Barbados of its tree cover (used as firewood for the numerous boilers that refined sugarcane), changed the local climate, eliminated habitats for native fauna (especially birds and monkeys) and greatly reduced biodiversity on the island. Almost two-thirds of the island's remaining flora and fauna are now made up of alien species brought in by the British, like sugarcane itself. Forest clearance removed the main source of soil nutrients (leaf litter), which led to declining fertility. Lacking protective vegetative cover, heavy seasonal rains soon carried topsoil away from the canefields – planted in regimented rows which accelerated runoff downhill – clogging up Bridgetown harbour with silt and hampering long-distance trade. As early as 1661, the President and Council of the island recorded that 'the land is much poorer, and makes much less sugar than heretofore'. There were some early attempts at soil conservation in Barbados, with animal manure being used to replenish nutrients on 'worn out' land. But as pastureland for livestock was at a premium on monocrop plantations, the usual European response to declining fertility was to search for fresh soils, shifting sugar production from island to island. Soil degradation became a widespread problem throughout the Caribbean, in British Jamaica, French Saint-Domingue and Spanish Cuba, and by the nineteenth and early twentieth centuries in new sugar-producing zones too, such as Australia, Hawaii, Java, the Philippines and Taiwan (developed by the Japanese to meet growing market demand there).

The economic success of the sugar enterprise – its profits helped to fuel the growth of the triangular Atlantic trading system – encouraged the extension of slave-based plantation methods to the production of other types of cash crops, with tobacco and cotton being two of the best documented. From the early seventeenth century, they became key export commodities from the colonial American South (and the region's natural wealth continued to be drained off even after the break with Britain). Tobacco cultivation dominated the economies of Virginia, Maryland and North Carolina, with exports rising in volume from around £50,000 in 1620 to over £30 million in 1700 as demand increased from smokers in Europe. In the Deep South – South Carolina, Georgia, Alabama, Mississippi and Louisiana – cotton emerged as the most important cash crop, especially after the advent of the Industrial Revolution. By the mid-nineteenth century, planters in the American South were producing two-thirds of the world's cotton supply. But the soils of the region generally had a low mineral content, while those of the Piedmont Plateau (rolling uplands which stretch from central Alabama through to Maryland) were also highly susceptible to erosion. Tobacco placed a heavy burden on nutrient-poor soils, requiring far more nitrogen and phosphorous than most crops, and cotton was only a little less demanding. Despite some limited attempts to sustain the fertility of the land, by rotating legume crops, applying animal manure and, by the late 1860s, experimenting with both guano and phosphate fertilisers, profit-motivated planters had few incentives to improve the condition of their soils when labour was expensive and fresh acres were cheap. When crop yields started to decline on nutrient-depleted soils, often within a few decades, many planters migrated westwards to cultivate new ground (Native Americans who refused to sell their homelands were forcibly removed to reservations). In 1819, one critical Euro-American planter dubbed the mining of the South's fertility a 'land-killing' system; the sheer abundance of cheap acres had encouraged the profligate use of the soil rather than prudent management. Not only did specialised monocropping soon exhaust stockpiles of nutrients accumulated over centuries, clearing the land for cultivation also resulted in the loss of between six and twelve inches of topsoil from the Piedmont Plateau, and accelerated erosion is still a significant problem in the region today.

North America provides another remarkable example of misunderstanding and misusing the soil. In the 1860s, as family farmers in New England began to run out of productive land, the fertile Great Plains were opened up to agriculture. Under the provisions of the US Government's Homestead Act (1862), any person who settled and 'improved' 160 acres of grassland – staying put for at least five years – was considered the rightful owner. But the practice of European-style agriculture in this unfamiliar semi-arid environment was to prove one of the worst 'ecological blunders' in world history. New railroads carried tens of thousands of Euro-American settlers to the Great Plains, signalling the end for the more ecologically attuned nomadic economies of indigenous peoples – Arapahos, Cheyennes, Commanches, Kiowas, Sioux and others – based on bison-hunting (see Chapter 2). Cattle and other domesticated livestock first occupied the bison's former range, and, from the late 1870s, the introduction of refrigeration technologies saw millions of pounds of beef being transported to consumers in both American and British cities every month. However, ranchers keen to cash in on the 'beef bonanza' overstocked the range, which was open to everyone during the early years of settlement, greatly reducing the available supply of grasses for forage. In 1884, it is estimated that some 5 million head of cattle were grazing the Great Plains north of Texas. During 1885–6, a dry summer diminished feed grass supplies further and a severe winter saw over 80 per cent of cattle perish, delivering a fatal blow to the businesses of many speculative ranchers, some of whom had stocked pastureland with four times the number of livestock it could reasonably be expected to support. But it was the growing demand for wheat on the world market that led to the biggest ecological calamity to befall the region: the Dust Bowl.

The vacuum created by the collapse of the 'cattle kingdom', which had imploded in only two short decades, was filled by family farmers from the East. The noted geologist-explorer John Wesley Powell, in his *Report on the Lands of the Arid Region of the United States* (1878), had warned that the desert-like Great Plains were ill suited to arable agriculture on account of insufficient rainfall and the region's fragile soils. However, pioneering farmers moving onto the plains embraced the more optimistic view that 'rain follows the plough'. According to this apocryphal theory, vigorously promoted by land speculators such as Nebraska's Charles Dana Wilber, cultivating crops and planting trees would increase the region's rainfall. And, at the beginning of the twentieth century, above-average rainfall on the Great Plains and wartime demand for wheat in Europe saw farmers enjoy bumper harvests and high profits. Between the years 1914 and 1919, around 11 million acres of virgin grassland in Kansas, Colorado, Nebraska, Oklahoma and Texas were ploughed up to plant wheat. Wartime profits were reinvested in new machinery to expand production, and by the 1920s small gasoline-powered tractors hauling ploughs to break new land were a familiar sight across the Midwest. But severe droughts are an 'inevitable fact of life' on the plains, recurring roughly once every twenty years. When prolonged drought hit in the 1930s, the light loess soils of the Great Plains, lacking the protective cover of native shortgrass vegetation, simply dried out and blew away. Hundreds of human-induced dust storms, some lasting for days, carried soil sediment as far as the cities of Chicago, New York and Washington, DC. In 1938, a US government official reported that the Great Plains were losing 850 million tons of topsoil every year to wind erosion. Millions of plains people abandoned their farms during the 'dirty thirties', and most of these 'exodusters' headed westwards to start over in California or the Pacific north-west. Nor was the American Dust Bowl an isolated case. Between the 1930s and 1950s, European 'environmental pioneers' in drought-prone regions of southern Africa, Australia, Canada and the Soviet Union also created dust-bowl conditions by applying monocrop farming methods in new landscapes that they did not fully understand.

Soil erosion surged for the third time after 1945, with the widespread adoption of heavy farm machinery, the intensification of global trade links and increasingly efficient food transportation networks by sea, land and air. In the 1850s, only 4 million tonnes of food had been traded around the world; by the late twentieth century this figure had reached almost 250 million tonnes. The third erosion surge was 'superimposed on its predecessors', and it affected both long-established agricultural areas in Europe as well as newly cultivated land in the wider world – especially in the tropics. Although agriculture in the USA was becoming mechanised by the 1920s, tractors and combines were rarely encountered in Europe before the Second World War.

The introduction of industrialised agriculture in Britain, and the growing size and weight of farm machinery, caused widespread soil compaction problems in the second half of the twentieth century. The compaction of topsoil can reduce microbiological activity and nitrogen levels, which inhibits plant growth. It slows moisture absorption too, heightening susceptibility to water erosion and flooding by increasing the volume of surface runoff after heavy rainfall. Since 1945, the uprooting of over 225,000 kilometres of field boundary hedges in England and Wales (some ancient and species-rich), to allow tractors and combines more room to manoeuvre, removed important barriers against water and wind erosion. Although rates of erosion are generally low in temperate Britain at between 3 and 11 tonnes per hectare per year, recent surveys show that 45 per cent of its arable land is now at risk of significant soil losses. Indeed, Britain's erosion rates greatly exceed those for soil formation of just 1 tonne per hectare per year. Deep compaction caused by farm machines, many weighing over 20 tonnes, is the most common form of soil degradation found across the European Union. It affected more than 33 million hectares of its cultivated land by the early 1990s, from south-eastern Britain, through the Benelux countries and northern France, down to south-western Spain, reducing crop yields by as much as 35 per cent. Soil compaction is also the main type of degradation in central and eastern Europe, where it now affects 62 million hectares or 11 per cent of the total land area.

After the Second World War, most agricultural expansion was concentrated in the tropical areas of the world, primarily at the expense of rainforests in central and west Africa, south and south-east Asia, and Central and South America. Developing countries, even after post-war decolonisation, continued to produce low-cost commercial monocrops for consumption in the developed world. Fast-growing markets in the USA, and the recovering economies of western Europe and Japan, fuelled demand. Political independence did not bring economic independence: unequal trading relationships still underpinned the global market system. Forest communities, whose subsistence techniques had been sustainable over centuries, were either displaced or turned into wage labourers as rainforests were cleared to raise cash crops for export. Intensifying economic integration, under American rather than European leadership, saw Ghanaian forests become cacao plantations, Malayan forests transformed into rubber plantations and Costa Rican forests cleared for pastureland to raise beef cattle. From the 1960s, forests in the Brazilian Amazon basin were opened up to agriculture, becoming a mosaic of cattle ranches and cultivated crops, producing everything from corn and coffee to sugarcane and soybeans. The conversion of tropical rainforests to large-scale agriculture played a key role in the third erosion surge. Deforestation exposed their shallow topsoils to the 'exceptional erosive power' of tropical downpours, resulting in severe land degradation. Moreover, as trees play a vital part in nutrient cycling in the tropics, once they were removed soils quickly became unproductive. Even grassland cannot be sustained for long, and nearly all the cattle ranches established in the Amazon basin during the 1960s and 1970s had been abandoned by the mid-1980s. As was the case in North America, frontier farmers moved on to clear new land as soon as the soil's fertility was exhausted. Since 1945, around 80 per cent of all significant soil degradation has taken place in the developing world, with the most serious impacts experienced in tropical countries.

To meet the rising demand for food and non-food crops (particularly cotton), with rapid global population growth once again a key driver, irrigation systems were massively expanded to boost production. Between 1950 and 2002, as Table 4.1 shows, the world's total irrigated land area almost tripled from 94 million to about 276 million hectares (or 18 per cent of all cropland).

At the same time as farm mechanisation quickened the pace of ploughing, sowing and harvesting, modern irrigation systems – by improving water supply – turned previously arid and semi-arid 'wastes' into fertile agricultural land, helping to increase crop production to unprecedented levels. By the turn of the twenty-first century, the roughly 18 per cent of cultivated land that is irrigated produced around 40 per cent of the world's food supply. Today, the largest areas of irrigated land are to be found in India, China, the former Soviet Union, the USA and Pakistan.

In the nineteenth and early twentieth centuries, Britain had led the world in irrigation science and engineering. The transformation of British India's irrigation from antiquated seasonal inundation canals to a modern perennial system, delivering a reliable year-round water supply, greatly increased agricultural productivity and the tax revenues of the colonial state. Hydraulic engineers viewed India's great rivers – the Ganges, Godavari, Indus and others – as natural resources to be conserved and controlled, so that their waters could be used profitably to raise irrigated crops, rather than simply 'running to waste' in the sea. The construction of barrages, canals, dams and

Table 4.1 The growth of global irrigated land area

Date	*Global Total (million hectares)*
1800	8
1900	48
1950	94
1960	137
1970	168
1980	211
1990	235
2002	276

Source: adapted from John McNeill, *Something New under the Sun: An Environmental History of the Twentieth Century* (London: Penguin, 2000).

weirs to create extensive river-fed perennial irrigation systems – most notably in the Punjab – led to considerable increases in crop yields, especially those destined for export such as wheat, cotton and sugarcane. Irrigation can extend the growing season, which meant that two or even three crops could be produced each year, instead of just one. The success of Indian irrigation schemes in boosting agricultural output from marginal semi-arid lands saw similar water-harvesting techniques spread throughout the British Empire and beyond. From the late nineteenth century, the ideas and innovations of hydraulic engineers in British India were taken up and developed further in Australia, Africa, Canada, the USA and elsewhere.

However, the construction of vast irrigation systems had a variety of adverse environmental side effects. Inadequate drainage led to waterlogging and salinisation, which both impaired plant growth and severely degraded the land. In India, salinity problems had reduced yields of Punjabi wheat as early as the 1860s. Between 1955 and 1985, persistent waterlogging and increasing soil salinity forced more than a quarter of India's irrigated land (13 million hectares) out of production. Rural workers faced a growing risk from waterborne diseases, particularly malaria. Irrigation ditches and over-saturated soils provided ideal breeding grounds for anopheles mosquitoes.

From the 1930s, numerous high dams were constructed around the world to extend irrigation acreage, to control floods and to provide hydroelectricity for urban and industrial use. At over 220 metres in height, the multipurpose Boulder (Hoover) Dam on the Colorado river in the USA, completed in 1935, was the first to inspire widespread imitation. As Richard White has shown, the Colorado river was remade and put to work as an 'organic machine'. Following the American model, some 40,000 large concrete dams had been built by the end of the twentieth century. Prestigious big-dam projects held particular appeal in newly independent countries and Communist states – places such as Egypt, China, India, Pakistan and the Soviet Union – to help 'legitimate governments and popularise leaders'. By the late twentieth century, about two-thirds of the world's river flow passed through networks of high dams. Silt and nutrients transported by great rivers were trapped behind these structures, reducing the fertility of floodplains and exacerbating coastal erosion. For example, since the completion of Egypt's Aswan High Dam in 1971, commissioned by its modernising leader Colonel Gamal Nasser, the delivery of the Nile's annual 'gift' of soil-renewing silt has virtually ceased. Further downriver, the Nile Delta is rapidly shrinking because insufficient sediment is available to keep erosion by the Mediterranean Sea at bay. And without the flushing action of the Nile's annual flood, which prevented salt build-up in irrigated soils, salinisation has emerged as a serious threat to Egyptian agriculture. To protect commercial cotton cultivation against the Nile's late-summer flood, the Aswan High Dam compromised the efficiency of what was for over five millennia an ecologically sustainable irrigation system. Today, between 30 to 40 per cent of Egypt's irrigated cropland is seriously affected by salinisation. Globally, recent estimates suggest that approximately 1.5 million hectares of the world's irrigated land is being lost annually to salinisation and waterlogging. These soil losses almost outstrip the rate at which nations can bring newly irrigated land into production.

Paradoxically, the problems associated with dams and irrigation can even lead to the spread of desertification (desert-like conditions). The Soviet Union provides the most dramatic example of this process. In the 1960s, the Aral Sea in Soviet Central Asia began to dry up, as hydraulic engineers diverted water from its two sources – the Amu-Darya and Syr-Darya rivers – mainly for the irrigation of cotton plantations in Uzbekistan. The Soviet government achieved its goal of becoming self-sufficient in cotton. However, deprived of its freshwater flow (reduced to a mere 10 per cent of its former influx), what had been the world's fourth largest lake lost more than 50 per cent of its area and 60 per cent of its volume. As the Aral Sea shrank, the local climate altered, many plants, animals and other organisms perished, its fishing industry collapsed, and its salinity more than tripled. The exposure of something like 45,000 square kilometres of its former lakebed created a salt-encrusted desert with frequent dust storms, which ruined crops and damaged human health. Communist agriculture in the twentieth century, with its emphasis on large-scale development, collectivism and economic growth, was no more environmentally friendly than the capitalist model.

Over the past half-century, around two-thirds of the world's agricultural land has been affected by the different forms of soil degradation outlined above. As demand for food and non-food crops continued to rise, and virgin land was in ever-shorter supply, public interest in protecting what was once seen as an inexhaustible resource – soil – began to increase.

Soil management and conservation

Soil-management techniques to combat degradation and to improve crop yields existed long before the twentieth century. Indeed, they have been embedded in many traditional forms of farming around the world for millennia. Although it is hard to generalise about different land-use practices in diverse cultures, common problems of preventing soil erosion and maintaining soil fertility did elicit some common solutions. Agricultural terracing is one of the main techniques that emerged to control erosion, particularly in mountainous areas of the world. An ancient anti-erosion measure, it involves building carefully spaced stone walls perpendicular to slopes in order to collect loose soil and to create planting platforms. Archaeological evidence suggests that there was no single point of origin for terrace construction. From around 4,000 years ago, it seems to have been developed independently by highland farmers in the Middle East, China, Europe and the Mediterranean, Africa, the Americas and elsewhere. Terracing not only retained rich topsoil, enabling the cultivation of steep hillsides, it also facilitated the irrigation of crops. One of its most important functions was to direct freshwater supplies from streams, springs and cloudbursts to cultivation platforms. In pre-Columbian Mexico, for example, terrace irrigation at Hierve el Agua in the Valley of Oaxaca supported continuous agriculture for at least eighteen centuries. Terracing continued to be built down into the modern period. In late-nineteenth-century Japan, for instance, engineers used narrow-based terraces to reduce erosion and to control flooding following the deforestation of its steep mountain slopes. Terrace construction accelerated worldwide from the 1930s, in response to growing concerns about soil losses following the American Dust Bowl. However, if terraces were not built robustly, properly maintained and correctly positioned, they could actually exacerbate erosion problems.

Pastureland also needed protection from the effects of overstocking and overgrazing. Transhumance, the seasonal movement of livestock, could spread the environmental impacts of grazing over considerable distances. By dispersing animal herds over large areas, transhumant pastoralism, which has been well documented in alpine and southern Europe, parts of Africa and the Indian Himalaya and northern plains, both reduced pressure on vegetation and minimised erosion from trampling. By frequently moving from one pasture to another, from lowlands to highlands, herders could avoid causing long-term damage to grazing areas. Over time, nomadic pastoralists developed an intimate knowledge of vegetation cycles and water resources on their migratory routes. They also formed workable and mutually beneficial relationships with other land users. Rice cultivators in India, for example, paid passing herders to have their animals manure the soil to enhance its productivity. The gains from such practices generally encouraged peaceful coexistence between sedentary farming and nomadic herding communities (but tensions inevitably surfaced from time to time). Although evidence exists for unrestrained resource use by some pastoralists, particularly in Mediterranean Europe, traditional transhumance is now viewed by many environmental historians as a relatively sustainable form of animal husbandry. Long-distance pastoralism, however, was unpopular with nineteenth- and twentieth-century colonial governments that aimed to foster more settled lifestyles among their subjects. It could also provoke clashes between nations where migratory routes crossed borders, such as the Sino-Indian conflict over access to grazing areas in the 1960s.

Mobility was crucial to protecting against erosion and maintaining soil fertility in traditional systems of 'swidden', 'shifting cultivation' or 'slash and burn' agriculture. Swidden farming had been employed the world over since the dawn of agriculture, particularly in tropical environments. In such systems, cultivators cut trees and undergrowth and then burnt them to provide fertilising ash. Initially, the nutrients released in the ashes boosted crop yields. But within a few years (often after just two or three crops), yields would fall, and farmers moved on to clear new plots of land. Several decades could pass before shifting cultivators returned, allowing vegetation to regenerate and soils to recover. Swidden systems essentially utilised a long-term fallow, and they are widely acknowledged as being among the most resilient forms of farming in world history. Commonly associated with intercropping (planting two or more crops in the same plot to improve food security), swidden was practised successfully by the indigenous peoples of the Americas, Africa, south-east Asia and elsewhere for millennia. Shifting cultivation remains an effective subsistence technique today where population densities are low and there is an abundance of land, such as traditional Amerindian agroforestry in the Amazon Basin.

To European eyes, the practice of shifting cultivation could seem primitive and wasteful, and a major cause of deforestation (see Chapter 3). In northern Europe, where populations were growing and fresh land was scarce, sedentary farmers had learned to maintain soil fertility by rotating the landscape through the farm. From the mid-sixteenth century, the new English practice of 'convertible husbandry' both retained soil nutrients and raised crop yields. Simply put, this agricultural system rotated pasture and arable land and fully integrated livestock into farming. Progressive farmers discovered that the key to improving soil productivity was to alternate crops and livestock on the same land. Previously, most land had been either permanent arable or permanent pasture. The classic Norfolk four-course rotation system (wheat, turnips, barley and clover), for example, was in common use by the eighteenth century. Food crops of wheat and barley were rotated with fodder crops of turnips and clover. Clover fixed nitrogen in the soil, while better-fed and more numerous cattle and sheep produced more manure for soil fertilisation. Although collecting and applying manure was a time-consuming and laborious task, by recycling animal wastes as fertiliser, good crop yields were sustained over time. Innovations in farm implements, better drainage and the utilisation of marl and lime to counteract soil acidity also helped to improve agricultural productivity. Wheat yields in England almost doubled, from twelve bushels per acre to twenty bushels, during its protracted 'Agricultural Revolution'. Similar practices had become standard across much of Europe by the nineteenth century, most notably in the Netherlands, France and Germany. Unproductive fallow periods were all but eliminated, and this contributed to a substantial rise in population. Between 1700 and 1800, Europe's population increased from roughly 81 million to around 123 million. These agricultural methods, designed to enrich the soil, were renewable in Europe (but they could be destructive when transferred to inappropriate environments overseas, as we have seen). In China, Japan and Korea, it was common for farmers to apply 'night soil' (human wastes) as well as animal dung as fertiliser. By structuring their operations to return nutrients and organic matter to the soil, pre-industrial farmers completed a 'virtuous circle' – what is now called a positive feedback loop – that enhanced its fertility.

Traditional agricultural practices relied heavily on natural processes to slowly improve soils and crop yields. But even in years of bountiful harvests, they rarely provided much above basic subsistence diets for most of the world's population. From the 1840s, however, the German scientist Justus von Liebig advocated taking a 'chemical approach' to the problem of agricultural production. Liebig had discovered that plant growth was limited mainly by nutrient depletion. He warned that increasingly intensive cultivation, particularly as practised in North America, was wearing out the land. According to Liebig's influential 'Law of the Minimum', if just one nutritive element was missing from the soil then crop yields would decline, even if others were present in abundance. He popularised the notion that crops could be grown continuously simply by adding nitrogen, phosphorous and other key nutrients directly to the soil. 'A time will come', Liebig predicted in 1843, 'when fields will be manured with a solution prepared in chemical manufactories.' While he acknowledged the value of organic manures and the cultivation of legumes in farming, Liebig made it clear to contemporaries that soils only needed to be 'chemically adjusted' to maintain their fertility. Soils began to be likened to machines; they sometimes needed fine-tuning to produce high crop yields.

The expansion of farmland over the past five centuries, and the development of new agricultural technologies and global markets in food, would not have sustained the huge rise in the world's population without heavy inputs of fertilisers. Until the nineteenth century, farms were largely self-sufficient in terms of producing manures and cultivating legumes to maintain soil fertility. The widespread use of Peruvian and Chilean guano – the nitrogen-rich droppings of seabirds – represented the first major shift from traditional forms of nutrient cycling. Between 1840 and 1880, Peru alone exported an estimated 12.7 million tons of guano (worth around £150 million) to farmers in Europe and North America. Although a natural source of fertiliser, commercial guano did not complete the 'virtuous circle' that saw nutrients recycled on the farm. Coming from outside the feedback loop, guano transformed the basis of farming, which increasingly involved the 'one-way transfer of nutrients to consumers'. Guano, which arrived fit for use in sacks from the suppliers, was less demanding on farmers' time and labour than collecting and applying animal manures. It also paved the way for the introduction of factory-produced chemical fertilisers by creating ready-made markets for 'artificial manures'.

Guano was expensive, and exploitable deposits were scarce. Growing anxiety that supplies of this natural resource would soon be exhausted stimulated research into chemical methods of fertilisation. The first chemical fertiliser was a concentrated 'superphosphate', produced by treating phosphate rock with sulphuric acid. Invented in 1842 by the English agronomist Sir John Bennet Lawes, the following year he began a long-running series of experiments to measure the effects of superphosphate and other fertilisers on crop yields at his Rothamsted estate (today the oldest agricultural research station in the world). Early test results on yields of winter wheat, spring barley and a number of other important crops were 'spectacular'. Liebig's vision of the agricultural future came closer to reality when Lawes opened a superphosphate factory in Deptford, London. Phosphate rock was soon being mined in Europe, the USA (after 1888), Morocco (after 1921), and by the Soviet Union on the Kola Peninsula (after 1930) to meet rising demand for inorganic fertilisers. But it was not until the development of the Haber-Bosch process to capture atmospheric nitrogen in 1909 that their use began to challenge the dominance of organic fertilisers. The growth of the chemical fertiliser industry in the first half of the twentieth century was slow, hampered by the intervention of two world wars and the Great Depression. However, as Table 4.2 shows, after 1950 the availability of relatively cheap nitrogen fertilisers, which supplied in abundance the plant nutrient that most often limited growth, saw a dramatic increase in their consumption.

Soil management began to rely on scientific expertise rather than local environmental knowledge, with the age-old problem of nutrient depletion apparently solved by the liberal application of the right minerals to the land at the right time, substantially increasing crop yields per hectare and maximising food and fibre production. The large-scale 'chemicalisation' of agriculture, particularly the use of nitrogen-based fertilisers, was the main reason that the world's population quadrupled in the twentieth century. Without the 'pumped-up' yields made possible by inorganic fertilisers, the global agricultural system could only support an estimated two out of three of today's 6.5 billion people.

Initially, the greatest consumption of artificial fertilisers was in Europe and North America, allowing farmers to take some marginal or degraded agricultural land out of production. Bigger harvests could be had from a smaller cropland area. From the 1960s, chemical fertiliser use began to expand rapidly in the Soviet Union, Japan, India, Mexico, the Philippines, China and elsewhere. Farmers no longer needed to depend on animal or green manures to maintain soil fertility, so traditional crop rotations and mixed-farming

Table 4.2 World inorganic fertiliser consumption

Date	*Inorganic Fertiliser (million tonnes)*
1900	2
1940	4
1950	10
1960	30
1990	150
2000	137

Source: adapted from Clive Ponting, *A New Green History of the World: The Environment and the Collapse of Great Civilisations* (London: Vintage, 2007).

systems could be dispensed with in favour of highly specialised monocultures. Post-Second World War food crops, like the high-yielding dwarf wheat and rice varieties adopted worldwide during the Green Revolution, were scientifically developed to be responsive to high doses of fertiliser, herbicides and pesticides. This fundamental change has seen a huge decrease in biodiversity in farmers' fields. Most of the world's food intake now comes from high-yielding varieties of just four crop plants: maize, potatoes, rice and wheat. And we should not forget that soil degradation was, and still is, commonplace on productive farmland. The intensive utilisation of inorganic fertilisers simply masked underlying problems of accelerating erosion, deep compaction and poor soil health. Their use actually disrupts the microbial activity that builds healthy soils. In addition, during heavy rains, chemical runoff from treated fields was carried into rivers, lakes, estuaries and coastal seas, which resulted in serious pollution problems such as eutrophication (water becomes too nutrient-rich and deoxygenated, causing dramatic algal blooms and fish mortality). During the past fifty years, the overuse of factory-produced fertilisers has altered cycles of nitrogen and phosphorous in the biosphere. The long-term ecological implications of interference with these nutrient flows remain unclear. However, recent efficiency campaigns aimed at farmers, together with the introduction of bioengineered crops that better resist pests and plant diseases, have helped to reduce agrochemical use from its peak 1990 level (see Table 4.2).

As far back as the mid-nineteenth century, attempts had been made by states to educate farmers to get the most from their soil. In 1862, for example, the United States Department of Agriculture (USDA) was established, and the US Government also funded the creation of land-grant colleges and universities that offered agricultural training for its citizens. Increased crop production was the main focus of these early initiatives, rather than soil conservation. But as evidence of soil erosion caused by poor farming practices began to mount by the late nineteenth and early twentieth centuries, revealed by scientific surveys undertaken in the USA, Russia and elsewhere, the emphasis began to change. The world's first national soil-protection organisation was set up in Iceland in 1907, as a response to severe erosion and desertification caused by overgrazing and deforestation. The most influential, however, was the American Soil Conservation Service, established in 1935 during the Dust Bowl disaster. It became the world's leading centre on soil research. Headed by respected soil scientists Hugh Hammond Bennett and Walter Clay Lowdermilk, the new Soil Conservation Service – an agency of the USDA – offered advice and assistance directly to farmers. Action programmes to conserve soils included the construction of terraces, contour tillage that followed the lie of the land (rather than ploughing straight up and down the slopes of fields), planting shelterbelts of trees, returning bare earth to grass and implementing effective grazing-management regimes. Its approach to conservation was pluralistic, and the methods employed to protect soils depended on local environmental conditions and the type of farming operation. Lowdermilk in particular was an enthusiastic advocate of terraces, which became emblematic of the battle against soil erosion, having seen them used successfully in both China and Japan. With the Soil Conservation Service's help, by the 1940s farmers on the Great Plains were again producing bumper crops of wheat. None of the above conservation techniques were innovative in themselves, but nationally coordinated programmes to protect the soil, viewed mainly as an inexhaustible and indestructible resource up until the Dust Bowl, were a new departure. It was an approach many other nations were to imitate.

Bennett and Lowdermilk were both great evangelists of the 'gospel of conservation', touring the world to promote soil-protection practices. They visited Canada, China, Mexico, South Africa and Venezuela among other places. In addition, between 1935 and 1951, more than 1,100 agricultural technicians from eighty-eight countries spent time at the US Soil Conservation Service to learn its methods. As was the case with scientific forestry, as the field became more professionalised, soil knowledge was also shared at international conferences and in research publications. In 1927, the first meeting of the International Society of Soil Science (renamed the International Union of Soil Sciences in 1998) was held in Washington, DC, and its journal *Soil Research* was first published the following year. Its second meeting was held in Leningrad in 1930. Six years later, the Soviet Government convened a pan-soviet conference in Moscow on controlling soil loss. The British Government organised an Imperial Agricultural Research Conference, held in London in 1927, to start to build empire-wide support for good soil stewardship. The outcomes included the founding of the Imperial Bureau of Soil Science (later renamed the Commonwealth Bureau of Soil Science), which between 1931 and 1962 published the important *Bibliography of Soil Science* triennially. A global overview of soil erosion was prepared by the Bureau's deputy director Graham Jacks and fellow soil scientist Robert Whyte, published in 1939 with the evocative title *The Rape of the Earth* (toned down to *Vanishing Lands* for its American readership). In it they stressed the role of the scientific expert in devising ways to tackle soil erosion and pinpointed southern Africa as one of the world's most degraded areas. But just as European farming methods did not necessarily transfer well overseas, neither did soil-conservation techniques that were untested in unfamiliar environments, as the case study below demonstrates.

Case study: Soil erosion and conservation in Basutoland, southern Africa

In the 1930s, Basutoland (modern Lesotho), a small and mountainous country roughly the size of Belgium, was a showcase for British soil-conservation programmes in Africa. The first national soil-conservation programme implemented in British Africa, it attracted visitors from around the continent and the world – including Hugh Hammond Bennett. Its designers were the head of Basutoland's Agriculture Department R. W. Thornton and his anti-erosion officer L. H. Collett (who had previously been sent on a study tour of the US soil-conservation programme). Begun in 1936, by the eve of independence in 1964 over 42,000 kilometres of terraces had been constructed, 2,500 kilometres of diversion furrows had been excavated, and 2,200 kilometres of meadow strips had been planted, leaving little of the country untouched. The primary aims of this massive conservation programme were to protect the landscape against erosion gullies (often called dongas), to preserve valuable topsoil and to improve water management in a land of climatic extremes (droughts and heavy rainfall were both common). But as it was rolled out, this nationwide project actually broadened the web of dongas across Basutoland, exacerbating its soil-erosion problems.

The predominant soil types found in Basutoland are 'duplex soils', characterised by a sharp difference in texture between a coarse, fertile upper layer of sandy loam and a finer, denser lower layer of clay. These duplex soils have a very low resistance to erosion, especially when left exposed to heavy summer rains. Fast-flowing runoff down steep slopes quickly removed the friable upper soil layer, resulting in the formation of gullies. From small beginnings, these gullies could grow into enormous chasms capable of swallowing roads and fields. In addition, the infiltration of the clay layer by percolating waters created internal erosion (soil piping or tunnelling) and instability underground, which eventually caused gullies when the surface collapsed – often unexpectedly. Indigenous Basotho agriculture had involved a minimal disturbance of native soils and vegetation. Until the 1860s, the Basotho – who were both herders and shifting cultivators – had access to an abundance of land. They grazed local Afrikander cattle, fat-tailed sheep and goats on highland pastures in the summer and lowland pastures in the winter, with livestock dispersed widely to avoid any excessive stressing of the landscape. When crops were planted, the earth was not left for very long in an uncovered condition. The use of wooden hoes limited the extent to which soil structure and microbial activity were disrupted. The broadcast sowing of seed (staple foods included sorghum, beans, and squashes), mimicking natural distribution patterns, prevented the erosive channelling of rainwater between regimented rows of crops. By employing a flexible system of shifting cultivation in an 'unlimited landscape' – a field was used for a few years and then taken out of production – cropland was given the opportunity to rejuvenate. Local customs tightly regulated the use of land, grasses, trees and shrubs, which meant that soil degradation was relatively rare in Basutoland before the arrival of Europeans in the early nineteenth century.

After a long-running series of armed conflicts with immigrant Boer settlers, in 1868 the Basotho had little choice but to ask the British Government for protection. The following year, the British negotiated a fixed borderline between Basutoland and the neighbouring Orange Free State. Under the terms of the Treaty of Aliwal North (1869), the Basotho forfeited large areas of their rich agricultural lowlands to the Boers. Confined within a much smaller space, and forced to farm on less productive hillsides, the Basotho's traditional land-use system no longer functioned as effectively. Moreover, new opportunities for trade altered indigenous farming practices that had previously been well attuned to the local environment. Interaction with European traders, settlers and missionaries radically transformed the Basotho's culture and economy, as grain and livestock were exchanged for cash and desirable consumer goods such as guns, metal tools, textiles, coffee, tea and sugar, as well as ploughs and horses. New crops and livestock were adopted as Basutoland became linked to regional and international markets, including wheat, Merino sheep and Angora goats. In the late 1860s, the discovery of diamonds in the Harts and Vaal rivers area (modern Northern Cape Province), and a global rise in wool prices, had promoted an economic boom in southern Africa. Traditional Basotho agriculture was rapidly and profoundly reoriented to provide food for mining communities and wool for export. In 1893, for example, Basotho farmers produced around 11,600 tons of wheat, 6,000 tons of maize and 500 tons of wool for market. But the introduction of plough agriculture, and the expansion of livestock numbers, had negative long-term consequences for soils that were highly sensitive to disturbance.

It is not surprising that by the 1880s the first reports by European observers of soil degradation in Basutoland had begun to emerge. The earliest references to accelerated erosion are to be found in the correspondence of missionaries, who often mention the existence of deep gullies or ravines scarring the landscape. By the early twentieth century, the 'overcrowding' of people and their livestock was causing significant soil loss around Basotho settlements (a problem created in no small part by the Europeans), with the erosive effects of 'heavy traffic' along roads and cattle tracks attracting most attention from colonial officials. Then drought and dust storms swept across the country during the 1930s, carrying away exposed topsoil and adding to official concerns (although wind erosion was less conspicuous in Basutoland than water erosion). The first reliable estimate of the amount of land affected by soil erosion is contained in Sir Alan Pim's and S. Milligan's report *The Financial and Economic Position of Basutoland* (1935), which noted that some 10 per cent of the landscape was 'threatened by existing dongas'. Milligan placed the blame for this situation squarely on the Basotho's lack of a 'sound hereditary knowledge of hill cultivation', and he recommended the immediate implementation of a nationally coordinated 'protective public works' scheme as the 'only hope for threatened areas'. However, rather than reduce soil loss, the conservation programme that Milligan prescribed was to accelerate the process.

The terracing of Basutoland's fields and pastures, intended to prevent serious scouring and erosion after torrential rains, was only successful when carefully planned and constructed. As Kate Showers has shown, the country's extensive terrace system was both badly built and poorly maintained, which led to the creation of gullies where none had existed before. Conservation works were not supervised by professional engineers on a day-to-day basis. To save money, inexperienced European foremen were hired to supervise 'native gangs' of labourers, who were often unwilling participants in the implementation of the national programme (some were paid only with food). Heavy rains caused serious damage to poorly positioned and inadequately built structures. Instead of slowing down and safely dispersing runoff water across the landscape, constant breaches of the terrace system channelled the erosive force of the rainwater downhill, stripping away friable topsoil and causing the formation of new dongas. Many Basotho farmers actively resisted the building of soil-conservation structures by the colonial state (as did indigenous people in other parts of southern Africa), not only because they were observed to aggravate erosion problems but also because they reduced the size of cropping areas and made fields more difficult to plough. However, keen to preserve their protective alliance with the British (necessary to ward off the threat of incorporation into the Union of South Africa), the Basotho were careful only to remove or modify terraces that were situated out of sight of colonial officials. While new soil-conservation structures were not the sole cause of gully formation in Basutoland, the colonial administration's assumption that terracing was a 'proven technology applicable to all landscapes' turned out to be mistaken. When environments were not properly understood, even conservation efforts could be counterproductive.

The programme's legacy was to be long-lasting. Despite many obvious deficiencies, post-independence soil-conservation projects in Lesotho continued to rely on similar structures for the rest of the twentieth century. Today, according to the United Nation's World Food Programme, Lesotho's degraded soils are only able to produce approximately 30 per cent of its national cereal requirements (food is now imported rather than exported), and its landscape is one of the most seriously eroded in Africa.

Conclusion

Agricultural societies have always had to deal with the dual problems of soil erosion and exhaustion. Although traditional land-use methods did not always result in sound stewardship of the soil, there is considerable historical evidence to show the long-term resilience of agricultural practices such as shifting cultivation and intercropping. Traditional agriculture often protected – even enriched – soils over centuries, as well as maintained some biological diversity in the system. But over the past half-millennium, the export of European agricultural practices and technologies, and the rising demands of Western consumer markets, have led to the widespread degradation of the world's soils. 'Environmental pioneers' employed techniques that were inappropriate to the place, often causing erosion rates to increase dramatically. Since the Second World War, when the use of machinery, agrochemicals, monocrop and irrigation systems escalated, boosting crop yields, soil health has continued to decline. Large-scale industrialised agriculture, increasingly overseen by scientific experts, has divorced soil productivity from the condition of the land. State-backed conservation networks were cooperating internationally by the late 1920s, but technical failures and the dominance of the chemical-industrial approach to soil fertility undermined attempts at arresting degradation processes. As nearly one-third of the world's cropland – some 1,500 million hectares – was abandoned over the past four decades, mainly due to poor soil management, we are now in danger of severely depleting a resource that cannot easily be replaced in human timescales. But, except for the dramatic 'Dust Bowl' episodes of the first half of the twentieth century, the degradation of the world's soils – and the ecosystem services that they provide – usually occurred too slowly and unobtrusively to attract much public attention.

By the late twentieth century, there had been a substantial increase in the size of farms and a great reduction in the numbers of agricultural workers (particularly in the developed world). By the new millennium, 'scientific' industrialised agriculture was feeding a much larger world population than traditional farming had a century before (although inequities in global food distribution meant that millions still went hungry). However, the intensive use of irrigation and chemicals in modern agriculture is no longer increasing crop yields, and the amount of available freshwater and suitable farmland is shrinking. Moreover, supplies of relatively cheap oil that support globalised industrial agriculture (it is used to produce fertilisers, to run farm machinery, to pump irrigation water and to process, pack and transport food) are fast diminishing. In the future, farming must become less oil dependent and more energy efficient. Between 1900 and 2000, global agricultural production increased six-fold, but this was only achieved by an eighty-fold increase in energy use. We now put far more energy into growing, harvesting and transporting food than we obtain from eating it.

Since the 1992 'Earth Summit' in Rio de Janeiro, there have been attempts to introduce worldwide guidelines for sustainable land management, such as the 1998 Tutzing proposal to create a convention for global soil protection, but no 'code of conduct' has been agreed. As awareness of the global implications of soil degradation grows, state-sponsored agricultural research and development (mainly undertaken in richer countries) has begun to focus on strategies to improve soil structure and to enhance biotic activity as well as to reduce erosion. They include new methods of intercropping to increase biodiversity and food security, more use of animal and green manures to build humus levels and improve water and nutrient retention capacities and the introduction of cover crops and no-till farming to avoid leaving soils bare and open to erosion. In recent years, many conservation schemes in the developing world have involved indigenous and poor village communities, acknowledging that their cooperation and time-tested knowledge of local environments can help to prevent soil loss. For example, the Government of Lesotho, in partnership with the United Nations International Fund for Agricultural Development, is now working together with local people on conservation-based farming, grazing management and projects to stabilise and reclaim the dongas that cover the country. Organic farming too is beginning to lose its recently acquired 'alternative' status, making a return to the mainstream as many farmers rediscover the importance of soil health to long-term agricultural productivity (and as health-conscious Western consumers demand chemical-free produce). But hybrid models of traditional and modern agriculture that could help to conserve and rebuild the world's soils are only slowly gaining ground.

Further reading

Robert C. Allen, 'The Nitrogen Hypothesis and the English Agricultural Revolution: A Biological Analysis', *The Journal of Economic History* (vol. 68, March 2008, pp. 182–204).

David Arnold and Ramachandra Guha (eds), *Nature, Culture, Imperialism: Essays on the Environmental History of South Asia* (New Delhi: Oxford University Press, 1995).

William Beinart and Peter Coates, *Environment and History: The Taming of Nature in the USA and South Africa* (London: Routledge, 1995).

William Beinart and Lotte Hughes, *Environment and Empire* (Oxford: Oxford University Press, 2007).

Geoff Cunfer, 'Manure Matters on the Great Plains Frontier', *Journal of Interdisciplinary History* (vol. 34, spring 2004, pp. 539–67).

Rohan D'Souza, 'Water in British India: The Making of a "Colonial Hydrology,"' *History Compass* (vol. 4, July 2006, pp. 621–8).

David Gilmartin, 'Scientific Empire and Imperial Science: Colonialism and Irrigation Technology in the Indus Basin', *Journal of Asian Studies* (vol. 53, November 1994, pp. 1127–49).

Rashid Hassan, Robert Scholes and Neville Ash (eds), *Millennium Ecosystem Assessment, Volume 1: Ecosystems and Human Well-Being – Current State and Trends* (Washington, DC: Island Press, 2005).

J. Donald Hughes, *An Environmental History of the World: Humankind's Changing Role in the Community of Life* (London: Routledge, 2001).

Shephard Krech, John McNeill and Carolyn Merchant (eds), *Encyclopedia of World Environmental History* (New York: Routledge, 2004).

James C. McCann, *Green Land, Brown Land, Black Land: An Environmental History of Africa, 1800–1990* (Oxford: James Currey, 1999).

John McNeill, *Something New under the Sun: An Environmental History of the Twentieth Century* (London: Allen Lane, 2000).

John McNeill and Verena Winiwarter, 'Breaking the Sod: Humankind, History and Soil', *Science* (vol. 304, November 2004, pp. 1627–9).

——(eds), *Soils and Societies: Perspectives from Environmental History* (Strond, Isle of Harris: The White Horse Press, 2006).

Robert B. Marks, *Tigers, Rice, Silk and Silt: Environment and Economy in Late Imperial South China* (Cambridge: Cambridge University Press, 1998).

Elinor G. K. Melville, *A Plague of Sheep: Environmental Consequences of the Conquest of Mexico* (Cambridge: Cambridge University Press, 1994).

David R. Montgomery, *Dirt: The Erosion of Civilizations* (Berkeley, Calif.: University of California Press, 2007).

Clive Ponting, *A New Green History of the World: The Environment and the Collapse of Great Civilisations* (London: Vintage, 2007).

John F. Richards, *The Unending Frontier: An Environmental History of the Early Modern World* (Berkeley, Calif.: University of California, 2003).

Kate B. Showers, *Imperial Gullies: Soil Erosion and Conservation in Lesotho* (Athens, Ga.: Ohio University Press, 2005).

Steven Stoll, *Larding the Lean Earth: Soil and Society in Nineteenth-Century America* (New York: Hill & Wang, 2002)

Ashbindu Singh, Thomas R. Loveland, Mark Ernste, Kimberley A. Giese, Rebecca L. Johnson, Jane S. Smith, John Hutchinson, Eugene Fosnight, H. Gyde Lund, Tejaswi Giri, Jane Barr, Eugene Apindi Ochieng and Audrey Ringler (eds), *One Planet, Many People: Atlas of Our Changing Environment* (United Nations Publications, 2005).

Ted Steinberg, *Down to Earth: Nature's Role in American History* (New York: Oxford University Press, 2002).

Richard P. Tucker, *Insatiable Appetite: The United States and the Ecological Degradation of the Tropical World* (Berkeley, Calif.: University of California Press, 2000).

B. L. Turner, William C. Clark, Robert W. Kates, John F. Richards, Jessica T. Mathews and William B. Meyer (eds), *The Earth as Transformed by Human Action* (Cambridge: Cambridge University Press, 1990).

Benno P. Warkentin (ed.), *Footsteps in the Soil: People and Ideas in Soil History* (Kidlington, Oxford: Elsevier, 2006).

David Watts, *The West Indies: Patterns of Development, Culture, and Environmental Change since 1492* (Cambridge: Cambridge University Press, 1987).

Richard White, *The Organic Machine: The Remaking of the Columbia River* (New York: Hill & Wang, 1995).

Donald Worster, *Dust Bowl: The Southern Plains in the 1930s* (New York: Oxford University Press, 1979).

Chapter 5

Cities and the environment

This chapter explores the environmental problems of the urban and industrial past. Every town and city is different, but they have always placed similar demands on the natural environment, and nature has always affected urban life. Ancient Rome suffered from air, land, water and noise pollution, from overcrowding and traffic congestion, from catastrophic fires and serious floods, as well as from the depletion of natural resources in its hinterland. Urban environmental problems did not begin with the rise of the modern city, but industrialisation – with its intensive use of fossil fuels – and rapid population growth extended their scale and scope dramatically. In 1800 there were just six cities in the world that had over 500,000 inhabitants: Istanbul, Tokyo, Beijing, Guangzhou, London and Paris. A century later there were forty-three cities that exceeded half a million in population, including sixteen of more than 1 million people, mainly located in western Europe, the Atlantic seaboard of North America and on the coastlines of colonies under European control. By the year 2000 some 388 cities had passed the million mark and sixteen megacities had topped 10 million. Hypercities with more than 20 million inhabitants, such as Mexico City, New York and Seoul-Injon, now house populations larger than entire countries (for example, more than Norway and Sweden combined). Spatially scattered, cities have never dominated the world map, and even today they cover less than 3 per cent of the earth's land area. But as major centres of trade, production and consumption, they have played a leading role in changing the face of the earth.

City growth required clean air, fresh water and plentiful food and fuel supplies for urban populations, as well as raw materials for both construction and industrial purposes. Initially, these resources were found within reach of new settlements. But over time big cities began to draw in resources from all parts of the globe, often impacting adversely on ecosystems that were far removed from urban sites of production and consumption. City authorities also needed to develop effective systems to remove the pollution and wastes they generated. From the second half of the nineteenth century, the engineering of extensive subterranean waste-disposal systems helped to create a new kind of environment; what Lewis Mumford called the 'underground city'. A focus on the metabolism of modern cities such as Manchester, the world's first real industrial city, reveals the long-term effects of urban living on the global environment, tracking their 'ecological footprints' far beyond their own hinterlands. It is an approach that shows the urban and natural worlds to be inextricably linked by resource flows and technological networks, breaking down the old binary opposition of the city and the countryside. With more than half of humanity (about 3.3 billion people) now living in fast-growing cities, their environmental impacts are likely to increase in the future. This chapter will also examine some of the main social and health consequences of worldwide urban and industrial development.

Early cities and the environment

The first cities were established in Mesopotamia, the Indus Valley and Egypt as irrigated agriculture emerged along the banks of the Tigris-Euphrates, Indus and Nile river systems (see Chapter 4). Water-control projects required a degree of social organisation and cooperation that encouraged the growth of urban centres. In addition, the increasing agricultural productivity of the Nile-Indus corridor was sufficient to support urban populations of several thousand people. Cities such as Uruk, Mohenjo-Daro and Memphis were religious, ceremonial and administrative centres, containing rulers, priests, scholars, technicians, bureaucrats, skilled craftsmen, soldiers, merchants and others. As urban settlements developed, levels of social stratification and occupational specialisation became more complex. Cities concentrated control over resources in the hands of monarchs, the aristocracy and high-ranking persons, and religious rituals – connecting the elite to the gods, and the people to the elite – sanctified the new urban order. Chinese civilisation grew up along the middle course of the Hwang-ho (Yellow River) during the Shang and Zhou dynasties, with water management, religious worship and military rivalry between local leaders among the driving forces behind the foundation of early settlements. Similarly, from about 1400 BCE, important ritual and administrative centres were established in Mesoamerica and South America, beginning with the large Olmec settlement at San Lorenzo in Mexico.

Early cities were very different in many respects from their modern counterparts, and not only because religion was at the heart of urban life. They were compact 'walking cities', with temples, public buildings and the marketplace all easily accessible on foot. Most were surrounded by substantial walls, which offered protection against 'predators of all kinds'. For example, archaeological excavations have indicated that the wall around the Shang capital of Ao (now Zhengzhou in north-eastern China), built of pounded earth c. 1600 BCE, stood approximately 10 metres high, averaged 20 metres wide at its base and had a perimeter of nearly 7 kilometres. It also protected the inhabitants of Ao against floods. Urban centres were home to wealthy elites, while the poor lived on the outskirts (the reverse of current residential patterns in most affluent Western cities). Many urbanites still worked in the fields during the day, returning to their homes within the city walls in the evening. An ancient ideal was that urban centres should be self-sufficient. For instance, one important aspect of Aristotle's concept of *autarkeia,* set out in his treatise *Politics* c. 350 BCE, was that cities should be independent of imports and exports. And the fortunes of the first urban centres were closely tied to the countryside around them. Adjoining fields, pastures and forests were the main source of their food, fibre and fuel supplies. Indeed, down into the Middle Ages many cities retained fields and orchards within their walls in order to better withstand a military siege. But this ideal of urban self-sufficiency was never actually attained.

City-dwellers, with their growing needs and desires, constantly drew on natural resources from beyond the surrounding area. From the outset, cities engaged in long-distance trade to obtain resources and luxuries that could not be found locally. Sumerian traders, for example, used river transport and donkey caravans to exchange cereals, ceramics and woollen textiles for precious metals, lapis-lazuli gemstones and wood products from the Mediterranean region and across the Indian Ocean via contacts in the Persian Gulf. Classical-period Athens, unable to feed its burgeoning population from its own agricultural land, had to import about half its grain from its colonies around the Mediterranean and Black Sea. Ancient Rome, the archetypal imperial city, imported a wide range of goods through supply networks that spanned Italy, the Empire and the wider world. Both water-borne and road transport brought, for example, silks, cottons and spices from China and India; grain from Egypt, Tunisia and Sicily; olive oil from Libya and Spain; tin and lead from Britain; and timber from the Alps and the Lebanon. Marble for the monumental architecture that made the emperor's power visible was shipped to Rome from Carrara in Tuscany, but it was also imported from Algeria, Turkey and the Greek isles. Ivory products and large numbers of wild animals used for sport in the Coliseum mainly came from eastern and northern Africa, causing local extinctions of some species. These early cities exerted increasing pressure on the lands that supported them and caused environmental problems – deforestation, soil erosion and species loss – in places far and near.

Given the importance of access to natural resources, it comes as no surprise to find that most early cities developed in coastal and riverside locations. Cities sited near the sea or other large bodies of water gained the advantage of good trade and transport links, as well as opportunities for fishing to help boost food supplies. Locale was also significant from the perspective of urban health. The influential Hippocratic treatise *Airs, Waters, Places,* written c. 400 BCE, stressed the importance of air quality, pure water and a salubrious setting when choosing settlement sites. Tainted air (miasma) and stagnant water from marshy ground, for example, were identified as potent sources of disease. The actual layout of ancient cities, however, owed less to ideas about how the environment affected health and more to their economic and religious functions. Although some cities grew haphazardly, usually around a readily defensible position, many were consciously planned. Grid plans were extensively used, reflecting both religious symbolism (sacred places dominated at the centre) and the need for effective communications. The first urban centres in the Nile-Indus corridor were built according to a regular grid plan, as were early Chinese cities. In Greece, Hippodamus of Miletus applied this type of design to Athens' port of Piraeus, and his blueprint was subsequently used in planning many other Hellenistic and Roman cities. The orderly Romans perhaps did most to disseminate the 'chequerboard' planning model, imposing a uniform grid pattern on the new cities that they founded at the frontiers of empire, from Timgad in Algeria to Winchester in the British Isles. Since antiquity, planners such as Hippodamus have aimed to create well-designed, harmonious cities. But the dual pressures of economic and population growth meant that – except in a very few cities – ideal patterns of urban design soon fragmented and broke down.

Some early cities grew to a considerable size, especially capitals, with Athens containing an estimated 200,000 people in 430 BCE, while at its peak in 150 CE Rome had more than 1 million residents. Overcrowding, inefficient waste disposal, vermin infestations and increased trade, travel and troop movements all made big urban centres vulnerable to outbreaks of deadly diseases. For example, according to Thucydides, the 'great plague' that struck Athens during the Peloponnesian War, killing up to 35 per cent of its population and permanently weakening the city-state, had its origins in Ethiopia. Because of the severity of population losses from exposure to infectious diseases (smallpox, bubonic plague, measles, influenza, typhus, tuberculosis and others), early cities only grew through a constant influx of rural immigrants. This has led some contemporary historians such as Clive Ponting to characterise ancient capitals as 'parasites' that drained the surrounding countryside of labour as well as natural resources.

More often, however, cities are represented as the embodiment of early civilisations, standing in stark contrast to wild, untamed nature and rural life that was culturally limited. Yet urban dwellers experienced a wide range of serious environmental problems. These are perhaps best documented in imperial Rome, then the largest city in the world. The Roman poet Horace (65 BCE–8 CE), for example, complained about its noise and smoke pollution and disapproved of suburban encroachment on fertile farmland around the city. There were frequent fires and floods, as many houses in the crowded city had wooden roofs and the Tiber river regularly burst its banks after heavy rains or snowmelt. Records show that there were twenty-seven major floods in Rome between 300 BCE and 200 CE (a rate of one every nineteen years). During floods, waste matter in Rome's main sewer – the Cloaca Maxima –

often backed up and inundated low-lying areas of the city. Refuse was a considerable problem, due to the tendency of Rome's citizens to dispose of their household waste in any convenient spot. Growing refuse heaps outside the city walls not only compromised people's health but also Rome's defences. The extensive use of lead for domestic water pipes, kitchen utensils, pottery glazes, children's toys and as a sweetener of wine (lead arsenate) undoubtedly caused ill-health among Romans – particularly the upper classes. The harmful effects of long-term exposure included impaired fertility and neurological damage. It has been suggested that lead poisoning weakened Rome's populace generally as well as the ability of its rulers to wield power, which precipitated the end of empire. Overall, data from Artic ice-core studies have shown that imperial Rome increased the release of lead into the environment by a factor of ten.

The first cities caused significant ecological changes, and urban environmental problems allied to the overexploitation of resources in their hinterlands contributed to their decline and, in some cases, fall. By 600 CE, Rome contained only about 50,000 people. Overuse of the soil brought about the collapse of the earliest urban centres in the southern Mesopotamian region of Sumer and the Indus valley (see Chapter 4). Most ancient cities failed to adapt harmoniously to their surroundings, although their environmental impacts were mainly local and regional in scale, rather than global.

Cities in the era of the Columbian exchange

After 1492, European maritime commerce extended around the globe, with its great merchant cities at the centre of intercontinental trade networks. Europe's leading commercial cities, such as Lisbon, Seville, Venice, Antwerp, Genoa, Amsterdam and London, competed with each other for prime position. The opening up of the New World was to significantly increase the global traffic in food crops, other commercially important plants, and animals, as Table 5.1 illustrates.

Columbus's 1492 voyage initiated a dramatic reordering of the world's ecology and economy, as contact was re-established between the previously isolated Old and New Worlds. European mariners and traders turned the oceans into 'intercontinental highways', linked by great port cities, hastening the progress of what the environmental historian Alfred W. Crosby has called the 'Columbian exchange'. This exchange of organisms, both macro and micro, brought different ecosystems together as never before: the world became a place without 'biological borders'. Iberian commercial ambition – the search for a new trade route to Asia – had been the catalyst for this uneven two-way process that was to transform global agriculture. Livestock and food plants were carried by European ships around the world. Wheat fields, sugar plantations, cattle ranching and sheep herding soon spread throughout Spanish, Portuguese and other European colonies in the New

Table 5.1 The Columbian exchange (selected items)

Americas to Eurasia and Africa	*Eurasia and Africa to Americas*
Food Crops	*Food Crops*
Avocado	Apple, fig, plum
Beans (kidney, etc.)	Asian rice
Blueberry	Banana, plantain
Cacao	Barley
Cashew	Citrus fruits
Guava	Coffee
Maize (corn)	Lentil
Manioc (cassava)	Mango
Papaya	Melon
Passion fruit	Millet
Peanut, pecan	Oats
Peppers	Olive
Pineapple	Onion, leek
Potato	Pea, chickpea
Pumpkin	Rye
Quinoa	Sorghum
Squashes	Sugarcane
Sweet potato	Wheat
Tomato	Wine grape
Vanilla	Yam
Wild rice	Root vegetables (beet, carrot etc.)
Other Plants	*Other Plants*
Quinine	Flax
Sunflower	
Tobacco	
Animals	*Animals*
Guinea pig	Cattle, oxen
Muscovy duck	Chicken
Turkey	Honeybee
	Horse, donkey
	Pig
	Sheep, goat
Diseases	*Diseases*
Syphilis	Bubonic plague
	Chickenpox
	Cholera
	Influenza
	Malaria
	Measles
	Mumps
	Scarlet fever
	Smallpox
	Yellow fever

Source: adapted from Geoffrey Symcox and Blair Sullivan, *Christopher Columbus and the Enterprise of the Indies* (Boston, Mass.: Bedford, 2005).

World. Where return passages are concerned, very few American animal species were domesticated for food. But from the sixteenth century New World staples such as maize, manioc and potatoes were widely adopted by farmers in Africa and Eurasia. For instance, European traders brought sweet potatoes and maize to China in the mid-1550s, and they quickly became primary crops in many areas. Commercial crops such as tobacco were also introduced to China, south-east Asia and Africa. European enterprise provided the impetus that brought coastal communities everywhere into communication, helping to diffuse highly productive new crops worldwide and providing the nutritional foundations for sustained population growth – not least of all in cities.

Port cities were the hubs of an embryonic global market economy, through which the natural goods extracted by Europe's seaborne empires flowed. Urban entrepreneurs, often with state support, founded banks and provided investment to encourage long-distance trade. By the mid-seventeenth century, for example, Amsterdam had become the 'warehouse of the world'. Its leading trading companies, such as the United East India Company (founded in 1602), the Northern Whaling Company (founded in 1614), and the Dutch West India Company (founded in 1621), imported, processed and re-exported everything from plantation-grown tobacco and sugar to silk and whale oil. Under the terms of the Treaty of The Hague (1669), the Dutch had even gained control over Norway's timber trade. Amsterdam had around 200,000 residents in 1700, and it was one of the world's richest cities. Indeed, the success of its overseas trading empire saw the Netherlands become the world's first nation with more than 10 per cent of its population living in urban centres (increasing to over 30 per cent by the late seventeenth century). However, as discussed in the earlier chapters of this book, new systems of commodity production and rising metropolitan demand for natural resources could have serious social and environmental consequences at the peripheries of empire, exemplified by the slave plantations associated with the Atlantic triangular trade.

Europe's expanding capitalist system was centred on its largest cities (often, but not always, capital cities), with most being located on major rivers or near the sea. Moreover, the rise of many colonial settlements was closely connected to the economic growth and development of 'world cities' such as Seville, Amsterdam, London and Paris. For example, ports such as Havana (1515, Spanish), New Amsterdam (1625, Dutch, now New York), Boston (1630, English) and Montreal (1642, French) were all creations of the Atlantic trade in furs, sugar, slaves, timber and tobacco. But as conduits for large numbers of people as well as natural goods, seaports and harbours were also gateways for infectious 'crowd' diseases. Syphilis is thought to have arrived in Europe in 1493 on Columbus's return from the Americas, borne by his ships' crews (although the origins of this disease are still contested). However, the unification of disease pools that followed the 'Columbian exchange' did not significantly raise European mortality rates. In contrast, Old World infections – most notably smallpox, measles and influenza – spread from such 'seeding points' to first decimate Amerindian populations, and later, following the 'Cook exchange', Australian Aborigine, New Zealand Maori and Pacific island populations. Non-Eurasian peoples had no genetic resistance or acquired immunity to disease-causing microbes transferred from Europe, Asia and Africa via new sea routes. European exploration, trade and settlement inadvertently caused what has been dubbed the 'Great Dying', with epidemic outbreaks of smallpox and other diseases bringing about a drop in native populations of as much as 90 per cent (see Chapter 1).

The mining of minerals, particularly the silver and gold that lubricated long-distance trade, also generated urban development. Many mining settlements were established in the New World, a seemingly limitless source of precious metals, and their growth tended to be rapid and unplanned. In 1545, the discovery of vast deposits of silver ore at Potosí, located in present-day Bolivia, saw this Spanish colonial mining outpost turn into a boomtown. By 1570 it was the largest urban centre in the Americas, with a population of some 150,000 inhabitants. Potosí produced around 60 per cent of the world's silver in the mid-sixteenth century, and the Spanish peso became accepted currency in every major port city. The Columbian exchange redirected wealth to Europe (where silver and gold were scarce). The hazards of the job, including rockfalls, silicosis and mercury poisoning from refining the ore, took a heavy toll on Native American miners, who comprised both wage-paid workers and forced labour draftees (*mitayos*). It has been estimated that as many as seven out of ten Amerindians who worked in the Potosí mines lost their lives. Large-scale mining in Central and South America had severe environmental consequences too: forests were consumed to make the charcoal required in refining silver and gold; the mercury used in the extraction process polluted air, land and water; and changes in regional hydrology caused by dams and deforestation resulted in disastrous floods. The total amount of mercury released into the environment has been recently calculated at 196,000 tonnes (between 60 to 65 per cent of which was emitted into the atmosphere). Overall, just as imperial Rome increased lead concentrations, research indicates that early modern Spanish mining operations in Latin America were responsible for significantly raising background levels of mercury in the global environment.

The flow of silver and gold from the New World to the Old was funneled through colonial ports. Between 1503 and 1660, more than 32 million pounds of silver and 360,000 pounds of gold were exported. Regular sailings by Spanish fleets from ports such as Cartagena in Colombia and Veracruz in Mexico brought the bulk of this wealth to Seville. Spanish monarchs spent much of it on costly military campaigns that aimed to unify their lands. American silver did not stay long in Spain, as it was also used to settle the

Habsburgs' substantial debts around Europe. In addition, about 20 per cent of this bullion was shipped directly across the Pacific via a new trade route, from the port of Acapulco in Mexico to the Spanish entrepôt of Manila in the Philippines, to finance trading enterprises in the Far East.

During the three centuries after 1500, China and India were the world's most productive economies. By 1775, Asia accounted for around 80 per cent of all the goods manufactured globally. In particular, Chinese silks and ceramics, and Indian cotton textiles, were highly prized in Europe and elsewhere. As few European goods found a ready market in Asia, silver from the New World was generally used in payment. India received huge amounts of silver from Europe and passed some on to China (which enjoyed a trade surplus with most nations). In total, about 75 per cent of all American silver was to end up in China. Japanese silver also wound up there, exported to acquire silks and other manufactured goods. China and Japan contained several of the world's biggest cities in 1700, with Beijing and Edo both exceeding 650,000 inhabitants. However, these great capitals were more inwardly oriented than their Western counterparts, functioning mainly as the political and cultural centres of their respective realms. Commerce and trade were left to secondary cities, such as the ports of Guangzhou and Nagasaki. Similarly, in Mughal India seaports such as Bharuch and Surat were the chief trading centres, rather than land-locked Delhi. Despite the greater wealth of Asia, European cities emerged as the most active in the global trading system. Their geographical position provided an early advantage, with the Mediterranean the starting point of routes to the East, and the Atlantic the 'highway' to the Americas. Seas connected the world and, by virtue of their commercial dynamism (bank credit, insurance, and chartered trading companies) and maritime strength (navigational knowledge, innovative ship design, and superior naval firepower), outward-looking European merchant cities became the powerhouses of international trade.

Sixteenth-century Seville, then seventeenth-century Amsterdam, were in the vanguard of a commercial capitalism that created a truly global economy, incorporating the Americas for the first time. As the volume of seaborne trade grew – with Europeans increasingly acting as middlemen carrying everyone's merchandise – port cities became more prosperous and more populous places. Indeed, many colonial settlements were constructed from scratch in coastal areas and along the banks of navigable rivers, especially in the New World. The expansion of maritime trade, to borrow John and William McNeill's phrase, 'turned the world inside out', with most large cities oriented towards the oceans. But until the turn of the nineteenth century, only a small minority of people lived in cities (no more than 7 per cent of the world's population). Global economic growth between 1500 and 1800 was modest by today's standards (the best estimates suggest less than one quarter of 1 per cent annually). Even as overseas trade intensified, societies that were still mainly agrarian in character could only support a small number of big cities. Seville, Venice, Genoa, Amsterdam and others still relied heavily on the natural resources of surrounding hinterlands for their survival. Nevertheless, urban-led European enterprise radically reshaped the world's ecosystems, which often led to their long-term simplification and degradation.

The rise of industrial cities

From the late eighteenth century onwards, the rise of industrial cities ushered in a new phase of environmental change. Britain was the first nation to undergo rapid urban-industrial expansion, followed by northern Europe, the USA and Japan. Driven by the push of agricultural modernisation, and the pull of factory work opportunities, urbanisation rates accelerated sharply in Britain, with others in the industrialising world catching up by the early twentieth century. In 1851, Britain – the birthplace of the Industrial Revolution – became the first country to have more than half of its population living in cities. By 1900, Germany and the Netherlands had also passed 50 per cent, and France was close to the 'half urban' mark at 45 per cent. But the USA did not reach this level of urbanisation until 1920, and Japan later still in 1935. The world's ten largest cities in 1900 were London, New York, Paris, Berlin, Chicago, Vienna, St Peterburg, Philadelphia, Manchester, and Tokyo (the only representative from Asia, reflecting a shift of economic power to the West). While capital cities still tended to dominate economic activity, the Industrial Revolution had created major urban centres and agglomerations such as Manchester, Chicago and the Ruhrgebiet that rivalled them in size – a significant break with the past. The total number of urban dwellers worldwide grew from 27 million in 1800 to 225 million in 1900, a greater than eightfold increase. Urban-industrial growth was to place ever-increasing demands on natural resources, including fossil fuels, causing local and global environmental problems to escalate.

Expanding cities sucked in raw materials and agricultural products from farther and farther afield. London was the world's largest city during the industrial age, and its development had been dramatic. In 1750 London had a population of 675,000, which had increased to nearly 2.4 million in 1850 and over 6.4 million by 1900. As the hub of Britain's free-trade empire (restrictive mercantilist practices were abandoned in the 1840s), most of the nation's imports and exports passed through the port of London. In addition, it was a leading centre of consumption and production. London's soaring population created a huge demand for mass consumer goods. Its industries included everything from shipbuilding and machine-making, through clothing and printing, to food-processing plants and breweries. In 1865, the British economist W. Stanley Jevons vividly portrayed just how important resources drawn from distant lands had become in maintaining urban-industrial growth:

> [U]nfettered commerce has made the several quarters of the globe our willing tributaries.The plains of North America and Russia are our corn fields; Chicago and Odessa our granaries; Canada and the Baltic are our timber-forests; Australasia contains our sheep farms, and in Argentina and on the western prairies of North America are our herds of oxen; Peru sends her silver, and the gold of South Africa and Australia flows to London; the Hindus and the Chinese grow tea for us, and our coffee, sugar, and spice plantations are all in the Indies. Spain and France are our vineyards, and the Mediterranean our fruit garden; and our cotton grounds, which for so long have occupied the Southern United States, are now being extended everywhere in the warm regions of the earth.

By the late nineteenth century, improved transport and communications (steamships, transcontinental railways, refrigeration technologies, and undersea telegraph cables) even permitted large quantities of perishable foods to be moved over long distances. In return, Britain's manufactures – such as cotton textiles from Manchester, woollen textiles from Leeds, lace from Nottingham, linens from Belfast, guns from Birmingham and steel cutlery from Sheffield – were dispatched to markets across the globe. Imperialism was still beneficial to Britain during the Industrial Revolution. Manufacturing was the prerogative of the metropolitan centre, while colonies were expected to supply food and raw materials cheaply, as well as provide vital outlets for finished goods. Unequal trading relationships that first began to emerge in the sixteenth century underpinned urban-industrial development in Britain, and later in northern Europe.

The 'ecological footprints' of cities – the land and water areas required to produce the resources that they consumed and to absorb the wastes that they generated – became bigger during the Industrial Revolution, particularly in Europe and the USA. The classic study of the growing demands that a nineteenth-century industrial city could place on the environment is William Cronon's *Nature's Metropolis* (1991). Cronon traced the enormous flows of natural resources such as wood, wheat and livestock into Chicago for processing in its sawmills, grain elevators and slaughterhouses, and then on to urban markets elsewhere by rail and refrigerated boxcar. He also highlighted the ecological consequences of the city's economic expansion. Former buffalo plains to the west of Chicago first became rangelands for cattle, and later Euro-American farmers ploughed under their native grasses and replaced them with monocultural fields of grain. White pine forests to the north in Michigan and Wisconsin were clear-cut, leaving behind treeless wastelands. By the 1890s, Chicago's success in dominating the timber trade meant that wood was entering the city from across the entire nation. Cronon's history emphasises the interconnections between the city and the countryside, reminding us that they are not separate entities. His work, however, has little to say about pollution problems in the urban environment.

Cities cannot function effectively without inputs of fresh water and clean air, as well as an efficient means of waste disposal. As industrial cities began to grow rapidly during the nineteenth century, the natural systems that supported urban life started to deteriorate. In Britain, for example, population densities had reached approximately 138,000 per square mile in Liverpool, 100,000 per square mile in Manchester, 87,000 per square mile in Leeds and 50,000 per square mile in London by the 1840s. Pure water was at a premium, as pressure on resources from both householders and businesses increased. Initially, there was a heavy reliance on nearby rivers, streams and well-water to meet local needs, with demand soon threatening to outstrip supply. In crowded Liverpool it was reported that access to fresh drinking water was so inadequate in poor neighbourhoods that working-class people had to 'beg or steal it'. Many businesses, particularly textile factories, coal-mines, chemical works, paper mills and iron and steel foundries, were prodigious consumers of water, with some firms using millions of gallons every day. After use, they discharged their effluents directly into local rivers and streams, treating watercourses as 'nature's drains'. Mistakenly, many contemporaries believed that rivers had the capacity to safely dilute dangerous wastes and to purify themselves. However, as new industrial towns and cities developed, both the quantity and quality of available water supplies began to decline markedly. The environmental consequences included the die-offs of fish populations and aquatic vegetation, as well as river water becoming so obviously polluted as to be unfit for drinking and industrial use (which acted as a brake on economic growth). By mid-century, no fish could survive in London's lower Thames, while the river Aire at Leeds contained 'refuse from water closets, cesspools, privies, common drains, dung-hill drainings, infirmary refuse, wastes from slaughter houses, chemical soap, gas [works], dyehouses, and manufactures, coloured by blue and black dye, pig manure, old urine wash; there were dead animals, vegetable substances and occasionally a decomposed human body.'

Urbanites falling into Victorian Britain's polluted rivers, John Hassan has written, risked death through poisoning rather than drowning. In other fast-urbanising and industrialising nations the situation was similar, with rivers such as the Rhine and Ruhr in Germany, the Allegheny and Passaic in the USA and the Watarase in Japan all experiencing severe pollution problems. The Rhine represents a classic case of transboundary pollution, as chemical and other harmful substances flowed across political borders – from Germany through France and the Netherlands – to the North Sea.

Efforts to improve water services were driven not solely by commercial interests but also by public-health concerns. In the nineteenth century, industrial cities were extremely unhealthy places, with overcrowding and environmental pollution combining to create ideal conditions for the transmission of epidemic diseases. Although the links between dirt and disease were not properly understood until the 1880s, anxieties about the spread of cholera, typhoid and other waterborne infections in Europe and the USA culminated in the large-scale construction of new water-supply systems. At the same time as impressive rail networks were making it easier to feed cities, complex and extensive subterranean water systems slaked their thirst. Britain – the first industrial nation – was a pioneer in piping fresh water over long distances to urban homes and businesses. For instance, by 1857, the city of Liverpool was supplied with water collected some 36 miles away at Rivington Pike, at a cost of over £1.3 million. Because of the crucial importance of pure water to the urban metabolism, and the expense of providing it, from the 1840s local councils, rather than private companies, tended to assume responsibility for improving supplies. In addition, the technology of water filtration began to be used to ensure the purity of supplies. The first sand-filtering system for a public water supply had been introduced in Paisley, Scotland, as early as 1804. By the turn of the twentieth century, major cities such as London and Liverpool in Britain, Hamburg and Berlin in Germany, and Poughkeepsie and Pittsburgh in the USA all employed sand filters to protect their drinking water. Anti-pollution legislation was passed to protect watercourses, such as the British Rivers Pollution Prevention Act (1876) and the Massachusetts Act to Protect the Purity of Inland Water (1886). But these laws were rarely drafted and implemented effectively, as neither city nor state authorities had any desire to hinder urban-industrial growth. Indeed, according to Franz-Josef Brüggemeier, in Germany's Ruhrgebiet, state policy was consciously designed to protect developing industries at nature's expense. There were few sanctions to prevent businesses pouring untreated wastewater back into the region's rivers. The health of cities, then, depended more on sound engineering than tough legislation.

Along with urban waterworks, engineers and municipal authorities also cooperated in the construction of vast sewerage systems to dispose of human wastes. During the nineteenth century, the failure of rudimentary urban sanitary services – first in Britain and northern Europe and then in the USA – to cope with the rising volume of sewage became a cause for public concern. Traditional cesspools and privies overflowed in densely populated industrial cities, and drains intended to deal only with surface water quickly became choked with filth (Japanese cities were generally cleaner, as householders carefully collected their wastes for sale to farmers). At a time when miasmatic ideas about disease causation still held sway, the offensive smells and stinks of the city were regarded as a real health hazard. In his highly influential *Report on the Sanitary Condition of the Labouring Population of Great Britain* (1842), the noted sanitarian Edwin Chadwick advocated the building of new sewage systems to improve urban health. Chadwick believed that 'all smell is disease', as did other proponents of miasma theory, and he stressed the need to dispose of malodorous matter before it could contaminate city air. Chadwick, and many municipalities, hoped to defray some of the costs of construction by selling recycled urban wastes to the nation's farmers as manure. However, cities produced far more sewage than could be utilised on the land, and, in any case, some farmers had come to prefer imported guano or factory-produced chemical fertilisers (see Chapter 4). Despite some successful experiments with sewage farming, by the mid-1870s the consensus was that urban wastes were simply an unhealthy nuisance that should be swiftly and effectively removed, rather than a potential 'mine of wealth'.

Work on London's famous sewerage scheme began in 1858, following what the Victorian press called the 'Great Stink'. During the hot, dry summer months, water levels in the Thames had fallen, leaving behind on its banks the sewage of millions of people to 'seethe and ferment under a burning sun'. This resulted in a stench so foul that Parliament was forced to adjourn for a week. Fearing an outbreak of epidemic disease, the Government gave the go-ahead to a project that eventually built around 1,100 miles of new sewers and diverted 31 billion gallons of sewage away from the city centre every year. After London had completed its sewer system, many other British towns and cities emulated its example. Evidence that good progress was being made is furnished by the Doulton company, which by 1880 was manufacturing approximately 3,000 miles of sewer and drain pipes annually. Colonial cities such as Singapore, Hong Kong and Manila acquired municipal-run sanitation infrastructure, at least in districts that were occupied by Europeans. British sanitary ideas were also adopted in the USA, and, by 1905, over 8,700 miles of sewers had been constructed in the nation's large cities. In Germany's Ruhrgebiet, the Emscher river was legally turned into an open-air sewage canal. Re-engineered for purpose – its bed was straightened and lined with concrete – by 1904 it received the wastes of about 1.5 million people annually, as well as industrial effluents from hundreds of mines and factories. The construction of Tokyo's sewer system had begun in 1884, and until 1922 its wastewater was discharged directly into Tokyo Bay. Removing wastes quickly – still believed to be a good idea after the advent of germ theory from the 1880s – significantly reduced the danger of contracting so-called 'filth diseases' such as typhoid, improving the health of urban populations and allowing industrial cities to grow. By the 1920s, most big cities in the industrialising world provided their residents with safe drinking water and efficient sewer systems to flush away wastes (business and middle-class areas usually enjoyed the best services). However, sewage treatment was neglected, with little thought given to environmental impacts beyond the discharge point. Urban sewage systems had not solved the problem of river pollution. They simply relocated wastes further downstream, often endangering neighbouring communities before ultimately draining into the sea. Between 1920 and 1950, however, sewage treatment plants do start to become commonplace in western Europe and the USA. Even so, as late as 1970, only 20 per cent of Japan's population lived in districts served by sewage treatment plants.

Sanitary reformers also sought to deal with the perennial problem of refuse generated in urban areas. Heaps of garbage – food scraps, animal droppings, ashes and other solid wastes – accumulated in crowded towns and cities. As late as 1900, some 10 million tons of droppings from horse-drawn traffic alone were estimated to fall on Britain's streets. As part of environmental strategies to prevent disease outbreaks, main thoroughfares were systematically cleaned and refuse was regularly collected. In the New York borough of Manhattan, for example, scavenging teams were collecting an average of 612 tons of garbage every day by the turn of the twentieth century. But there was no consensus among sanitary authorities on the best method of disposal. Carting rubbish to great dumps outside of the city was expensive, although some solid wastes could then be recycled and reused. The poor found employment there as dumpsite scavengers and 'trash pickers'. Horse manure was sold to local farmers, as well as organic garbage as pig feed, while ashes from stoves and furnaces had value when converted into building materials. By the late nineteenth century, municipal destructors (incinerators) were considered a hygienic and efficient way of removing refuse. Britain led the way in the development of refuse destructors, and, by 1912, there were over 330 in operation in its large towns and cities. A similar number (approximately 300) were operating in the USA and Canada in 1914. However, incinerating plants elicited numerous complaints about dust and unpleasant odours, as well as the profligate waste of recyclable resources.

Some cities simply chose to dump their garbage into the sea. The British city of Liverpool employed two steamers, the *Alpha* and *Beta,* to dispose of its refuse in the Irish Sea. New York also had garbage scows that deposited their contents in the waters beyond the city's harbour. Dumping at sea drew complaints from fishermen whose nets were fouled and also from coastal communities as garbage washed up on beaches. In the long term, the rising tide of urban refuse was to be hidden away in 'sanitary landfills'. Garbage was tipped into deep trenches and covered over with a layer of earth every evening, putting the problem 'out of sight and out of mind' for most city dwellers.

By the early decades of the twentieth century, most industrial cities had enjoyed a measure of success in 'solving' the problems of water supply and of sewage and refuse disposal (largely by displacing them elsewhere). In contrast, a solution to the 'smoke nuisance' in urban areas remained elusive. Inexpensive bituminous coal was the key fuel of the Industrial Revolution, with steam power gradually outstripping muscle, wood, wind and water power as the main source of manufacturing energy. World coal output was around 10 million tons in 1800, with Britain producing more than 80 per cent of the total. By 1900, global coal output had risen somewhere between eighty-fold and a hundredfold, with both Germany and the USA now major producers. In Japan, the Hanshin region between Kyoto, Kobe and Osaka had become the biggest Asian zone of heavy industry. Coal smoke, rather than decomposing organic matter, was now the main urban air-pollution problem. The forests of tall chimneys that dominated the skylines of industrial cities such as Manchester, Essen, Pittsburgh, and Osaka, some over 100 metres in height, were designed to reduce local air pollution by discharging smoke far up into the atmosphere to be dispersed by the prevailing winds. It was erroneously thought that the earth's atmosphere was an inexhaustible sink that could dilute and ultimately neutralise all the pollutants, including smoke, particulate matter (soot), sulphur dioxide and carbon dioxide, which industrial chimneys could emit. For example, in 1901, the eminent German chemist Clemens A. Winkler had insisted, 'The volumes of consumed coal disappear without a trace in the vast sea of air.' However, coal smoke did not disappear as harmlessly as contemporaries hoped.

Topographical and meteorological conditions often prevented the dispersal of coal smoke away from industrial cities. Manchester, for instance, bounded by the Pennine chain of hills, found that its factory smoke became trapped for days at a time during spells of cold, calm weather. When tall chimney-stacks did function effectively, they simply displaced smoke pollution to plague other communities situated downwind. Moreover, as supplies of fuelwood declined, millions of urbanites began to burn coal in their homes for heating and cooking (although wood and charcoal were still crucial to Japanese householders well into the twentieth century). Domestic smoke emissions, released into the air at street level, were difficult to disperse even when strong winds were blowing. Coal smoke came to characterise the industrial city, and it caused a variety of social and environmental problems.

Smoke pollution – by absorbing and scattering light – was recognised to lower sunshine levels significantly in industrial towns and cities. It blocked out, reformers complained, as much as 50 per cent of available sunlight and daylight. The inhabitants of coal-fuelled cities lived in a permanent smoke haze, which by the late nineteenth century was generating grave concerns about declining urban health and morals. Rickets, a bone-deforming disease caused by lack of sunlight and poor diet, was endemic in urban populations on both sides of the Atlantic (but not in Japan where vitamin D from a diet that contained plenty of fish enhanced the absorption of calcium). Smoke pollution was also closely associated with rising death rates from respiratory diseases such as bronchitis, particularly during prolonged 'smog episodes' when air quality was very low. The term 'smog' was coined in 1905 by Dr Harold Des Voeux of the London-based Coal Smoke Abatement Society to describe the fusion of smoke and fog. Western city-dwellers were not only thought to be degenerating physically but also morally, as crime, gambling, drunkenness and disorder were all believed to flourish beneath the thickening smoke cloud. The situation was exacerbated by the retreat of the prosperous middle classes to suburbs situated upwind of the smoke, leaving the working classes behind with few appropriate 'role models' to emulate.

While the health risks and moral dangers associated with coal smoke were distributed unequally between the classes, the labour burdens imposed by this form of environmental pollution were divided unequally along gender lines. The all-pervasive smoke and soot filtered through the narrowest cracks and fissures to soil everything within the home. House cleaning and washing clothes were costly, time-consuming and strenuous activities that locked legions of women, in their roles as housewives and domestic servants, into a never-ending round of drudgery. In early twentieth-century Essen, it was reported that homes needed cleaning at least twice a day. At the same time, the economic costs of cleaning in Pittsburgh were estimated to be between 50 and 75 per cent higher than if the city had been free from smoke. And Monday, the day on which women lit substantial domestic fires to procure the hot water necessary to carry out the weekly wash, was – somewhat ironically – the smokiest day of the week in many British cities. Coal smoke affected the daily lives of all urbanites to some extent, but exposure to air pollution and experiences of the problems it caused varied considerably according to class and gender.

Smoke pollution, denser in winter than in summer, seriously damaged architecture and vegetation in urban-industrial areas. Monumental public buildings, such as new town halls that expressed civic pride, were soon defaced by soot and grime. Acid rain – the term was neologised in 1872 by the British scientist Robert Angus Smith – caused the stonework of historic buildings, such as the Houses of Parliament, St Paul's Cathedral and York Minster, to crumble. In the USA and Germany, smoke, soot and ash had similar deleterious effects on the built environment. Chicago's dazzling 'White City', an exhibit of elegant buildings created for the 1893 World's Fair by some of America's best architects, quickly darkened and decayed, while in 1929 the opening of a new coal-fired powerplant at Herne in the Ruhrgebiet left much of the town covered with ash. 'Smoke City Osaka', as it was known to the Japanese, experienced roughly the same levels of soot and grime as European and American manufacturing centres. The 'smoke nuisance' was also responsible for blighting the few green spaces that remained in fast-growing industrial cities. Reduced light intensity, heavy sootfalls and acid rain meant that trees, shrubs and flowering plants all struggled to survive. By the 1880s, parks and gardens in urban-industrial areas began to be dominated by flora proven to be capable of enduring the smoke. Among the most common species to be found in British cities were poplar trees, privet, elder, ash, hawthorn, honeysuckle, lupins, rhododendrons and willows. The harmful effects of air pollution could also be seen right across industrial regions as the toxic emissions of factories, smelters and chemical works damaged great swathes of vegetation between towns and cities, including forests and crops.

Anti-smoke organisations, which were springing up in both Britain and the USA by the turn of the twentieth century, argued that clean air was as important to urban health as pure water and fresh food. Pressure groups such as London's Coal Smoke Abatement Society (founded in 1898) and

Chicago's Anti-Smoke League (founded in 1908), called for the strict regulation of smoking chimneys to reduce pollution. But early smoke-control laws, like those relating to river pollution, were rarely implemented effectively in industrial cities reluctant to place unnecessary constraints on economic growth. Despite growing awareness of the damage caused to buildings, vegetation and the people's health, there was very little popular support for the smoke-abatement movement. Most urbanites associated smoking industrial and domestic chimneys with progress, employment and prosperity, rather than aesthetic loss, declining biodiversity or a waste of finite resources and human life. In Germany's Ruhrgebiet and Japan's Hanshin region, 'powerful nationalism' also underpinned a willingness to endure high levels of atmospheric pollution. Not until a series of well-publicised 'killer smogs' occurred in the Meuse Valley, Belgium (1930), Donora, Pennsylvania (1948) and London (1952) did public opinion begin to change. Smog episodes were belatedly recognised to be as hazardous to health as epidemic outbreaks of cholera and typhoid had been in the nineteenth century. Tougher legislation was introduced, such as Britain's Clean Air Act of 1956, which outlawed both industrial and domestic smoke emissions in urban areas. In addition, the expansion of cleaner energy systems – gas, electricity and oil – made a major contribution to clearing the skies of coal smoke. Air quality in the cities of the first industrial nations began to improve significantly from the 1950s onwards. However, explosive urban-industrial growth in other parts of the world during the second half of the twentieth century was to increase the demands that cities made on water, land and air.

Cities and the environment after 1950

Most affluent cities in western Europe, the USA and Japan had made considerable progress in improving local and regional environmental conditions by the 1960s and 1970s, first by curbing sanitary and then smoke pollution problems. During the past half-century, however, urbanisation and industrialisation in Latin America, the Soviet Union, Africa, China and south Asia saw the same issues reoccur, as well as the emergence of environmental pollution problems that are truly global in scale.

The world as a whole experienced rising urbanisation levels after 1950, as Table 5.2 demonstrates, particularly in developing countries. While urban growth began to slow down in the developed world, cities in the developing world, such as São Paulo, Mexico City, Lagos, Manila, Mumbai, Seoul and Shanghai, expanded phenomenally. Since 1990, the urban population of developing countries has increased by an average of 3 million people every week. In 1950, there were eighty-six cities with a population that numbered over 1 million; by the year 2000 there were 388, and sixteen megacities had more than 10 million. This growing concentration of population,

Table 5.2 Changing levels of urbanisation, 1910–90 (percentage of total population)

Region	*1910*	*1950*	*1990*
Western Europe	45	63	78
United States	46	64	75
Japan	40	56	77
Latin America	7	41	71
Soviet Union	14	39	66
Africa	5	15	34
China	5	11	33
South Asia	8	16	28
World	18	29	43

Source: adapted from John McNeill, *Something New under the Sun: An Environmental History of the Twentieth Century* (London: Penguin, 2000).

consumption and production in urban areas has greatly increased the stress placed on the environment.

Meeting the increasing urban demand for food and fibre has had a major impact on forests, soils, rivers and wildlife (see previous chapters). Cities in developing countries, when growth was not properly planned, also struggled to deliver safe water supplies and effective sanitation and waste-disposal systems. As was the case in Victorian cities, population growth often outstripped the capability of basic urban infrastructure to keep pace. Indeed, the scale and scope of post-Second World War development has dwarfed the nineteenth and early twentieth-century experience. In 1990, around 800 million city-dwellers in the developing world still lived without piped water, sewage systems or access to a proper toilet. Because human wastes fouled the streets, or were discharged directly into rivers used to supply drinking water, life-threatening 'filth diseases' such as typhoid fever and gastro-intestinal disorders were still common in the cities of the developing world. Untreated wastewater from burgeoning industries increased the severity of river pollution. Inadequate refuse disposal was (and is) a major health hazard too. For example, in Freetown, Sierra Leone, only around half of the solid wastes that the city produces are collected and processed by the municipality; the rest is dumped in back streets and along roadsides, or in bodies of water or storm drainage channels. In many African, Asian and Latin American cities, the poorest slum-dwellers (particularly women and children) find employment as scavengers on noxious garbage dumps, recycling wastes such as glass, plastics and metals into reusable products. Unhealthy, densely packed slums are found in most cities of the developing world, with the upper classes often living in outlying suburbs or protected 'gated communities', such as São Paulo's Alphaville (defended by more than 800 private guards). Environmental inequalities are highly visible, and the lack of basic sanitary services and poor hygiene in slum areas currently accounts for around 1.6 million deaths per year globally.

Smoke pollution continued to be a major problem, particularly in the cities of the Soviet Union, eastern Europe and China where industrialisation was still heavily dependent on coal. The commitment of Communist countries to economic growth – to match or even out-produce the capitalist West – resulted in 'polluted skies as never before'. Monopolistic state control over heavy industries such as coalmines, chemical works and iron and steel foundries meant that there was little pressure to introduce abatement technologies or environmental protection legislation that might impede economic expansion. While smoke emissions were coming under control in the West, they were rising rapidly behind both the Iron and Bamboo Curtains. In Nizhni Tagil, an industrial city 700 miles east of Moscow, urbanites regularly experienced 'night at noon' because the smog was so thick. Chinese industrial cities, such as Shenyang and Lanzhou, were also characterised by high levels of smoke pollution. China, which now has over 120 cities with more than 1 million inhabitants, currently burns in excess of 2 billion tonnes of coal per annum (and it is likely to remain its dominant fuel for decades to come). In 1998, of the ten most polluted cities in the world, nine were to be found in China. Lanzhou's inhabitants had to breathe air with average levels of pollution that were more than 100 times the World Health Organisation's (WHO) guidelines. Located at the bottom of a narrow river valley, Chinese city planners unsuccessfully attempted to solve Lanzhou's pollution problem by blasting the tops off the surrounding hills to allow the smoke to escape. In 2006, the WHO estimated that over 1.5 million people around the world were killed annually by respiratory and other diseases associated with air pollution (the great majority in developing countries).

While smoke pollution had declined in the developed world by the 1970s, the problem of acid rain persisted and spread. Legislation designed to control visible coal smoke did little to curb invisible emissions of sulphur dioxide, one of the main components of acid rain, largely because preventive flue-gas scrubbing systems were prohibitively expensive. Worried that low-level concentrations of sulphur dioxide posed a threat to human health, regulators instead insisted on another 'technical fix' – raising the height of industrial chimneys – to better disperse and dilute this harmful pollutant. Coal-fired power stations, providing 'clean energy' in the form of electricity to homes and industry, produced much of the sulphur dioxide that reacted with moisture in the atmosphere to form acid rain (solving one environmental problem can often create another). By 1960, Britain had built more than sixty new power stations and greatly extended the generating capacity of many older installations, with their chimneys reaching heights in excess of 135 metres. These tall chimneystacks, intended to reduce local air pollution, transported sulphur emissions over hundreds and even thousands of kilometres. By the end of the 1960s, Scandinavian scientists had shown that enormous flows of air pollution from Britain, carried by the prevailing winds, were causing lakes and rivers to acidify in Norway and Sweden. The ecological consequences also included the widespread decline of forests (although acid rain is just one of a number of cumulative stresses that can cause die-offs). Though there had been regional acid rain problems since the onset of the Industrial Revolution, they intensified and assumed international proportions. Industrial sulphur-dioxide emissions drifted unhindered across national borders, and this type of transboundary pollution also became an issue elsewhere in Europe and the wider world. South-westerly winds carried air pollution from Germany, Czechoslovakia and Poland to Scandinavia. By the mid-1980s, about half of Canada's annual sulphur deposition was found to originate in the USA. And from the 1990s, Japan was regularly showered by acidic rainfall from China and Korea. International cooperation, such as the implementation of the United Nations Economic Commission for Europe Convention on Long-Range Transboundary Pollution in 1979 (the USA and Canada signed up together with a large group of European countries), has begun to reduce acid rain by regulating sulphur-dioxide emissions from utilities and industry, but it still remains a problem today.

Mass car ownership after the Second World War saw unregulated exhaust emissions of nitrogen oxides increase dramatically, becoming a significant source of acid rain. Private cars, once lauded as the 'clean' alternative to horse-drawn transport, have also contributed to many other environmental problems. Automobiles encouraged urban sprawl, covering more land with bricks and concrete, as people no longer had to live within walking distance of their work or on a rail, tram or bus route. Until recently, vehicle exhausts spewed millions of tonnes of toxic lead into the atmosphere (tetraethyl lead was used as an additive in petrol to help the engine run smoothly), impairing the normal intellectual development of children in urban areas. Somewhat ironically, from the mid-1950s smoke-control initiatives in Western cities allowed more sunlight to penetrate to the streets, where it reacted with pollutants emitted from vehicle exhaust pipes to form dangerous ozone-laden photochemical smogs. Concerns about the effects of air pollution from automobiles on human health first emerged in Los Angeles – a city built for cars – as its residents began to complain of smarting eyes and a wide range of respiratory ailments. By the 1970s, Athens, Greece, experienced some of the worst traffic pollution in the world, and during prolonged photochemical smog episodes the city's death rates climbed by as much as 500 per cent. At the turn of the twenty-first century, there were around 600 million cars registered worldwide, 200 million of them in the USA alone. As car ownership has risen sharply in the developing world, places such as Bangkok, Buenos Aires, Mexico City and Mumbai now rank among the most heavily polluted 'smog cities'.

Since 1950, urban-industrial growth has contributed substantially to global climate change. The Intergovernmental Panel on Climate Change (IPCC), a scientific body set up by the World Meteorological Organisation and the United Nations Environment Programme, has reported that global temperature increase over the past fifty years has been unusually rapid when compared with that for the previous two millennia. While cities are not the only generators of greenhouse gases such as carbon dioxide, nitrous oxide and ozone, they undoubtedly produce more emissions than non-urban areas. Greenhouse gas emissions from homes, commercial buildings, industries and transport (including cars) are very high in urban areas, although levels vary considerably between cities according to lifestyles, incomes and energy use. Burning fossil fuels – coal, oil and natural gas – to meet our increasing energy needs accounts for over 60 per cent of all emissions. Wealthy cities in the developed world tend to produce more greenhouse gases than poor cities in the developing world. As economic activities have shifted from industries to services in cities such as Berlin, London, New York and Tokyo, most of their emissions now come from the energy consumed by commercial buildings, homes and transport. As Asian economies have returned to prominence, the industrial sector accounts for most greenhouse gas emissions in cities such as Guangzhou, Lanzhou, Shanghai, Kolkata and Mumbai, although their householders currently consume far less energy per capita than those in the West. China's citizens produce an average of around 5 tonnes of carbon dioxide each year, and India's less than 2 tonnes, while those in the USA produce over 19 tonnes per person annually. In 2007, however, industrialising China overtook the USA to become the leading emitter of greenhouse gases. Unless an effective global agreement can be brokered to greatly reduce carbon-dioxide emissions, then an estimated temperature rise of between 1.8 and 6.4 degrees Celsius by the end of the twenty-first century will see 'natural' disasters such as heatwaves, droughts and forest fires become more frequent and severe. Many of the world's coastal cities – more important than ever to international trade – will be placed at risk from rising sea levels caused by climate change. Global warming is already responsible for around 300,000 deaths per annum, and it brings disease, hunger and poverty to millions, mainly in developing countries.

Case study: nineteenth-century Manchester, the 'workshop of the world'

Manchester was the first modern industrial city, the place that pioneered not only the factory system but also the *laissez-faire* and free-trade principles that were to become predominant in Victorian Britain. The city was the centre of the cotton textile trade, and during the first four decades of the nineteenth century the Manchester region's products constituted over 40 per cent of the value of the nation's exports. Burgeoning demand for cotton goods, particularly ready-made clothing, in North America, Europe, Africa and Asia had stimulated the industry's phenomenal growth. Although cotton's share of the value of British exports had fallen to 25 per cent by 1913, it was still the country's largest export industry. The population of Cottonopolis, as Manchester was widely known, increased from nearly 77,000 inhabitants in 1801 to more than 316,000 in 1851 (Greater Manchester's population exceeded 2.1 million by 1901). The city's spectacular transformation from a largely verdant and countrified town into the archetypal industrial city generated a dynamic new image of Manchester as a powerhouse of productivity, progress and wealth creation, as well as negative responses to its overcrowded, unsanitary slums and polluted environment. Foreign visitors and travellers, such as Friedrich Engels, Leon Faucher and Alexis de Tocqueville, flocked to see futuristic Manchester, with the latter writing in 1835, 'From this foul drain, the greatest stream of human industry flows out to fertilise the whole world. From this filthy sewer, pure gold flows. Here humanity attains its most complete development and its most brutish; here civilisation works its miracles, and civilised man is turned back into a savage.'

Becoming the 'workshop of the world' not only created extremes of wealth and poverty, technological innovation and chaotic urban growth, it also increased the flows of natural resources into Manchester enormously, as well as its outputs of waste materials.

In the early nineteenth century, some 90 per cent of Britain's cotton industry was concentrated in the Manchester region (south-east Lancashire, north-west Derbyshire and north-east Cheshire). Raw cotton consumption in Britain grew from 5 million pounds in 1780 to an annual average of 937 million pounds by 1856, transforming the landscape of the American South where most of the crops were cultivated. Imports of raw cotton also came from Egypt and India after the American Civil War (1861–5), with similar environmental effects (see Chapter 4). By the turn of the nineteenth century, Manchester was no longer capable of feeding its own population. Only 25 per cent of its annual cereal requirements were sourced locally, although farmers in Lancashire and Cheshire played an important role in supplying the city with potatoes and other vegetables. In 1842, it was estimated that 50,000 tons of potatoes were sold in Manchester's markets every year. From the 1820s, live cattle, sheep and pigs were being imported from Ireland in large numbers via Liverpool. A more modest trade in German and Dutch livestock was carried on through Hull and other ports on the east coast. The advent of the railways, and the construction of the Manchester Ship Canal (36 miles from the sea, the city became the country's fourth-largest port), speeded the flow of both food and fibre into the city, and by the turn of the twentieth century supplies were coming in greater quantities from as far afield as Australasia, Asia, Africa and the Americas. Refrigerated meat, for example, was arriving from New Zealand, Australia and Argentina. In addition, Chinese and Indian tea sweetened with Caribbean sugar fortified the factory operatives of Cottonopolis to work long hours. By 1900, the ecological impacts of Manchester's escalating demands for food and fibre extended far beyond its immediate hinterland.

Manchester may have been the home of *laissez-faire,* but by 1851 its water supply was under municipal ownership. In 1845, the provision of clean piped water in the city was inadequate, with consumption levels of barely 2 million gallons per day. Access to fresh drinking water in the poorest districts of Manchester had been limited, with almost half of its citizens having to rely on unsafe sources of supply such as river water or wells situated next to overflowing graveyards and cesspools. Outbreaks of cholera in 1832 and 1849 and appalling sanitary conditions in slum areas encouraged Manchester's council to take its waterworks out of private hands. As well as concerns over deteriorating public health, the poor provision of water to the city's businesses also prompted this intervention. Numerous enterprises – factories, dyeworks and warehouses – were unable to obtain sufficient quantities of good-quality water to meet their day-to-day needs. To improve supplies to both domestic and industrial consumers, the city began to tap water sources in distant upland areas. Water was piped into Manchester from the Pennine hills, with the Longdendale chain of five reservoirs being the largest in the world when they were completed in 1877. The environment of Longdendale Valley was altered drastically as three hamlets and numerous farmsteads were submerged. Thirlmere Reservoir 100 miles north in the Lake District was added in 1894, despite opposition to the scheme from eminent nature preservationists such as John Ruskin, William Morris and Octavia Hill. One of the most famous beauty spots in Britain was dammed and converted into a reservoir to provide Mancunians with pure water and to assure the continued economic growth of the city. By 1900, piped water consumption in Manchester had leapt to almost 32 million gallons per day. However, the city's rivers – the Irwell, Irk and Medlock – were all badly contaminated by toxic wastes discharged from its industries, which had diversified to include chemical, engineering, printing and other works. The city council, which counted many industrialists among its members, did not rigorously enforce environmental-protection legislation, such as the Rivers Pollution Prevention Act (1876), that might have damaged the local economy.

If Manchester was ahead of most cities in providing a reliable supply of pure water, it lagged behind where sewage disposal was concerned. The city had a working sewer system, originally intended to deal with stormwater, but in 1868 only about 10,000 of Manchester's 67,000 homes had water closets that were connected to it (mainly in well-to-do neighbourhoods). The policy of the council at this time was to discourage the use of flushing toilets in order to ease the pollution burdens already placed on the city's rivers, as their waters were becoming unusable for industrial purposes. In 1866, for example, the Manchester Statistical Society reported that the river Medlock was so polluted by manufacturing wastes and domestic sewage that 'small birds have been seen to walk upon it.' Until the 1890s, the vast majority of Manchester's homes still used pail closets, ash pits, midden privies or cesspools, with some 70,000 tons of solid human wastes per annum being deposited at its Carrington Moss sewage farm, helping to feed the city. However, the cost of collecting and transporting this 'night soil' for use on the land was roughly twice its value as manure. More importantly, with pails, pits, middens and cesspools infrequently emptied in the poorer districts of the city, infant diarrhoea was still a major killer in late Victorian Manchester. High infant death rates from so-called 'summer diarrhoea', contracted by the faecal contamination of food and water, only started to decline significantly in the early decades of the twentieth century, after a comprehensive waterborne system of sewage disposal was provided for the city. Between 1890 and 1901 Manchester's existing network was extended to a total length of over 1,700 miles of sewers, including the construction of a new treatment works at the Davyhulme outfall on the river Irwell, situated to the south-west of the city. While liquid wastes were released directly into the river (not treated to a high standard for much of the twentieth century), many thousands of tons of sewage sludge were transported by barge down the Manchester Ship Canal to be dumped into a designated area of the Irish Sea (a practice that continued right up until 1998). After 1890, the steps taken to expedite the removal of the city's sewage undoubtedly saved lives, but they simply displaced this pollution problem to another part of the environment.

Rapid industrialisation and urbanisation saw a huge increase in energy consumption in Manchester. Coal mined in Lancashire, Staffordshire and Yorkshire fuelled the city's factories and heated the homes of its inhabitants, with its consumption rising sharply from just 100,000 tons in 1800 to some 3 million tons per annum in 1876. As Manchester's businessmen and householders rarely burned coal efficiently, smoke pollution became a serious environmental problem, blocking out the sun, destroying vegetation, defacing buildings and damaging people's health. As water supply and sanitary infrastructure gradually improved, reducing mortality from infectious 'filth' diseases considerably, the bronchitis group of respiratory disorders had become the biggest single killer in Manchester and other industrial cities by the turn of the twentieth century. The working-class district of Ancoats in Manchester recorded some of the country's highest levels of air pollution, as well as some of its highest death rates from respiratory diseases. Lack of sunlight was strongly linked to the spread of rickets among the city's children and to growing anxieties about the physical and moral degeneration of its citizens more generally. These concerns gained currency in 1899, when 8,000 of 11,000 Mancunians who volunteered for military service in the South African War were rejected as 'physically unfit to carry a rifle'. Further afield, acid rain caused by the sulphurous smoke emissions from Manchester's industrial chimneys – the city had sprouted around 1,200 tall smokestacks by 1898 – transformed the vegetation of the southern Pennine hills. Vast carpets of *Sphagnum* mosses, which had previously dominated in this upland locality, almost completely disappeared due to the extreme acidity of the peat. The cotton grass and bilberry moorland that prevails today replaced them. Ironically, a representative area of the changed moorland of the southern Pennines, its unique features acknowledged to be the result of the Industrial Revolution, has been preserved as a Grade 1 Site of Special Scientific Interest. Moreover, scientific research has clearly shown that many of Britain's lakes started to acidify in the mid-nineteenth century due to the increase in air pollution.

In January 1931, the city suffered a severe smog episode that resulted in 450 'excess deaths' from respiratory disorders. Out of this catastrophe emerged a campaign for the introduction of smokeless zones, organised by the National Smoke Abatement Society, which was based in Manchester. The campaign came to fruition in 1952, when local smoke controls were enforced in the city centre in advance of the national 1956 Clean Air Act. As the world's first industrial city, Manchester had always been in the vanguard of initiatives to deal with environmental problems caused by urban growth and entrepreneurial enterprise. The city's first anti-pollution pressure group, the Manchester Association for the Prevention of Smoke, was founded as early as 1842. It took more than a century of campaigning, however, before the smoke cloud that had enveloped the city finally began to clear. Indeed, progress towards improving urban environmental conditions in the city overall was slow and haphazard, often requiring a 'technical fix' that simply transferred pollution elsewhere. Today, exhaust emissions from Greater Manchester's 1 million or so cars are its biggest air-pollution concern, causing photochemical smogs during the summer months and contributing significantly to global warming. Private automobile use – 86 per cent of all journeys in Greater Manchester are made by car – adds significantly to its carbon-dioxide emissions, which are running at around 32 million tonnes per year. Although the city reinvented itself as a leading service provider (education, retail, finance, culture and transport, especially air travel) as most of its heavy industries migrated overseas, it is still the 'pollution capital' of Great Britain. Approximately 50 per cent of Manchester's households are affected by excessive noise; 25 per cent drink water supplied through lead pipes; 10 per cent of the city's land is heavily contaminated by heavy metals and other hazardous materials from manufacturing processes; water quality in the rivers Irwell, Irk and Medlock is still 'only slightly less than toxic'; and household waste is expanding at about 3 per cent annually. In 2004/5, Greater Manchester produced a total of 1.4 million tonnes of household wastes, most of which was disposed of in landfills outside of the conurbation's boundaries (only 16 per cent was recycled). Overall, the city-region's 'ecological footprint' is now roughly 125 times the size of its land area of 1,286 square kilometers. Despite recent attempts to project itself as a 'Green and Pleasant Region', Manchester still has some way to go in cleaning up its environment and achieving long-term sustainability.

Conclusion

The increasing concentration of the world's population in cities, particularly after the Industrial Revolution, has caused considerable environmental change. As the 'metabolic rates' of modern cities accelerated with growing production and consumption, their impacts on the environment extended from being mainly local and regional in scale to global. Food and other natural resources flowed into cities via complex transportation networks that criss-crossed the world (supermarkets today carry up to 50,000 different product lines), transforming the environments and economies of distant places. Great engineering projects – dams, pipelines and subterranean sewer systems – altered the hydrological landscape to supply urbanites with fresh water and to remove potentially hazardous human wastes quickly if not satisfactorily (they were usually displaced downstream). Rivers became 'nature's drains' as untreated industrial wastes were discharged directly into watercourses, while unwholesome refuse accumulated in dumps adjoining slum areas. Increasing fossil-fuel consumption by households and industries adversely affected urban air quality in the form of coal smoke and vehicle pollution, as well as contributed to transboundary and global-scale problems such as acid rain and climate change. Industrial civilisation did not set out to intentionally damage the environment, but its commitment to economic growth and development meant that the protection of nature was until recent decades low on the political agendas of most local and national governments.

It is clear that the poorest city dwellers have had to bear disproportionate pollution burdens, especially in terms of exposure to health hazards. While cities in the developed world have greatly improved their environmental conditions, many of those in the developing world are still struggling to provide adequate access to safe water supplies and sanitation for their poorest residents. Since 1950, urban growth rates have been highest in the developing world, especially in Asia, Africa and Latin America, although average resource consumption has been much lower per citizen than in the more affluent West. The United Nations Human Settlements Programme (UN-HABITAT) has estimated that the number of city dwellers in developing countries will more than double over the next four decades, from 2.3 billion in 2005 to 5.3 billion in 2050. Improving living standards and well-being for the world's poorest urban dwellers is one of the Millennium Development Goals adopted by the member states of the United Nations in 2000. Achieving this goal sustainably is a pressing challenge for the future. Globally, it is likely that about 70 per cent of the world's burgeoning population will live in cities by the middle of the twenty-first century.

However, it should be noted that cities do have the potential to allow humans to live more harmoniously with the natural world. In the early twentieth century, the Garden City movement championed the importance of planning to better integrate nature in urban space. Today, parks and gardens in some European cities support more biodiversity than the surrounding countryside. Carefully planned and effectively regulated cities, with compact forms, high population densities and economies of scale, also offer real opportunities to reduce resource use, pollution and waste. City living can help to meet the challenges of sustainable development through a wide range of initiatives, including maximising the use of renewable resources; minimising fossil-fuel use; constructing energy-efficient homes and businesses; recycling more refuse, organic wastes and wastewater; encouraging walking, cycling and the use of public-transport systems; and reducing the demand for land relative to population.

Green shoots of optimism are emerging in the developing world where urban planning is concerned. In 2008, for example, China signed a deal with Singapore to design and build a sustainable city – an eco-city for 350,000 people – on a site near Tianjin. Intended as a model for the future planning of cities in China and the wider world, Tianjin Eco-City aims to combine a high quality of life with a small ecological footprint by moving forward 'one credible step at a time'. All its buildings will be the 'last word' in energy efficiency, although initially only 20 per cent of its power will come from renewable resources. Lush green spaces will help to preserve local biodiversity (70 per cent of its plants will be indigenous species), absorb carbon dioxide and provide relaxing recreational areas. A compact city, it will be laid out to discourage driving, with 90 per cent of all trips being made on foot, cycle or by public transport (keeping its carbon-dioxide emissions low). More than 60 per cent of its wastes will be recycled and reused, including wastewater. All hazardous industrial and domestic wastes will be rendered non-toxic through treatment. The development of Tianjin Eco-City will be as environmentally friendly as possible without being overly idealistic (some critics have dismissed the eco-city concept as utopian). Even if the model is not widely replicated in future planning, the world's cities, as UN-HABITAT has observed, will be the 'front lines in the battle for sustainability'.

Further reading

William Beinart and Lotte Hughes, *Environment and Empire* (Oxford: Oxford University Press, 2007).

Franz-Josef Brüggemeier, 'A Nature Fit for Industry: The Environmental History of the Ruhr Basin, 1840–1990', *Environmental History Review* (vol. 18, spring 1994, pp. 35–54).

Mark Cioc, *The Rhine: An Eco-Biography, 1815–2000* (Seattle, Wash.: University of Washington Press, 2002).

William Cronon, *Nature's Metropolis: Chicago and the Great West* (New York: W. W. Norton, 1991).

Alfred Crosby, *The Columbian Exchange: Biological and Cultural Consequences of 1492,* 30th Anniversary edn (Westport, Conn.: Greenwood Press, 2003).

Devra Davis, *When Smoke Ran like Water: Tales of Environmental Deception and the Battle against Pollution* (Oxford: Perseus Press, 2002).

Mike Davis, *Planet of Slums* (London: Verso, 2006).

Ian Douglas, Rob Hodgson and Nigel Lawson, 'Industry, Environment and Health through 200 Years in Manchester', *Ecological Economics* (vol. 41, no. 2, 2002, pp. 235–55).

Angela Gugliotta, 'Class, Gender, and Coal Smoke: Gender Ideology and Environmental Injustice in Pittsburgh, 1868–1914', *Environmental History* (vol. 5, April 2000, pp. 165–93).

Christopher Hamlin, *Public Health and Social Justice in the Age of Chadwick* (Cambridge: Cambridge University Press, 1998).

John Hassan, *A History of Water in Modern England and Wales* (Manchester: Manchester University Press, 1998).

Rashid Hassan, Robert Scholes and Neville Ash (eds), *Millennium Ecosystem Assessment, Volume 1: Ecosystems and Human Well-Being – Current State and Trends* (Washington, DC: Island Press, 2005).

Alf Hornberg, J. R. McNeill and Joan Martinez-Alier (eds), *Rethinking Environmental History: World-System History and Global Environmental Change* (Lanham, Md.: Altamira, 2007).

J. Donald Hughes, *An Environmental History of the World: Humankind's Changing Role in the Community of Life* (London: Routledge, 2001).

Shephard Krech, John McNeill and Carolyn Merchant (eds), *Encyclopedia of World Environmental History* (New York: Routledge, 2004).

Bill Luckin, *Pollution and Control: A Social History of the Thames in the Nineteenth Century* (Bristol: Adam Hilger, 1986).

Tony McMichael, *Human Frontiers, Environments and Disease: Past Patterns, Uncertain Futures* (Cambridge: Cambridge University Press, 2001).

John McNeill, *Something New under the Sun: An Environmental History of the Twentieth Century* (London: Allen Lane, 2000).

John McNeill and William McNeill, *The Human Web: A Bird's-Eye View of World History* (New York: W. W. Norton, 2003).

Peter J. Marcotullio and Gordon McGranahan (eds), *Scaling Urban Environmental Challenges* (London: Earthscan, 2007).

Martin Melosi, *The Sanitary City: Urban Infrastructure in America from Colonial Times to the Present* (Baltimore, Md.: Johns Hopkins University Press, 2000).

——*Effluent America: Cities, Industry, Energy and the Environment* (Pittsburgh: University of Pittsburgh Press, 2001).

Carolyn Merchant, *Radical Ecology: The Search for a Livable World* (New York: Routledge, 2005).

Stephen Mosley, *The Chimney of the World: A History of Smoke Pollution in Victorian and Edwardian Manchester* (London: Routledge, 2008).

Lewis Mumford, *The City in History: Its Origins, Its Transformations, and Its Prospects* (London: Secker & Warburg, 1961).

Clive Ponting, *A New Green History of the World: The Environment and the Collapse of Great Civilisations* (London: Vintage, 2007).

Dieter Schott, Bill Luckin and Geneviève Massard-Guilbaud (eds), *Resources of the City: Contributions to an Environmental History of Modern Europe* (Aldershot: Ashgate, 2005).

Roger Scola, *Feeding the Victorian City: The Food Supply of Manchester, 1770–1870* (Manchester: Manchester University Press, 1992).

Richard Sennett, *Flesh and Stone: The Body and the City in Western Civilization* (New York: W. W. Norton, 1994).

John Rennie Short, *The Urban Order* (Oxford: Blackwell, 1996).

Vaclav Smil, *Energy in World History* (Boulder, Col.: Westview, 1994).

David Stradling, *Smokestacks and Progressives: Environmentalists, Engineers, and Air Quality in America, 1881–1951* (Baltimore, Md.: Johns Hopkins University Press, 1999).

Susan Strasser, *Waste and Want: A Social History of Trash* (New York: Metropolitan Books, 1999).

Eric Tagliacozzo, 'An Urban Ocean: Notes on the Historical Evolution of Coastal Cities in Greater Southeast Asia', *Journal of Urban History* (vol. 33, September 2007, pp. 911–32).

Joel A. Tarr, *The Search for the Ultimate Sink: Urban Pollution in Historical Perspective* (Akron, Ohio: University of Akron Press, 1996).

——'The Metabolism of the Industrial City', *Journal of Urban History* (vol. 28, July 2002, pp. 511–45).

B. L. Turner, William C. Clark, Robert W. Kates, John F. Richards, Jessica T. Mathews and William B. Meyer (eds), *The Earth as Transformed by Human Action* (Cambridge: Cambridge University Press, 1990).

UN-HABITAT, *State of the World's Cities 2008/09: Harmonious Cities* (London: Earthscan, 2008).

Anthony S. Wohl, *Endangered Lives: Public Health in Victorian Britain* (London: Methuen, 1983).

Chapter 6

Conclusion

Beyond the limits?

Many of the environmental problems that pose an enormous challenge to humanity now, and for the future, are deeply rooted in the past. From the fifteenth century onwards economic expansion, technological innovation, population growth and increasing urbanisation have seen humans transform the earth to an unprecedented degree. Overseas colonialism, especially the advance of European empires after 1492, reordered the world's ecology. Wholesale introductions of domesticated species such as wheat, barley, rice, cattle, sheep and goats by European colonisers created 'neo-Europes' in the Americas, Australia, New Zealand and elsewhere. The diffusion of high-yielding New World crops throughout the Old, together with plantation-grown coffee, tea, sugar and cacao, also improved food supplies and laid the foundations for a sustained rise in the world's population. But the spread of commercial enterprise and monocultural agriculture around the globe had devastating long-term environmental consequences, including deforestation, species loss and soil erosion. Natural resources were often 'mined' to exhaustion as traditional constraints on their use broke down at the frontiers of expanding European and non-European empires. The dominant attitude towards nature in early modern Europe was perhaps best expressed by the English philosopher and statesman Francis Bacon, who announced, 'The world is made for man, not man for the world.'

Since the Industrial Revolution, which Paul Crutzen favours as the starting point for the Anthropocene, ecological disruption has accelerated. From the late eighteenth century, great industrial cities and agglomerations such as Manchester, Pittsburgh and Germany's Ruhrgebiet began to emerge. New fossil-fuelled technologies were used to power factories, operate transportation networks and run people's homes. As places of mass production and mass consumption, modern cities required ever-increasing inputs of cheap energy, food and fibre, fresh water and other natural resources from an ever-widening hinterland. Evolving distribution networks – turnpike roads, canals, railways, steamships, electricity cables, water, gas and oil pipelines, container fleets, motorways and airports – allowed more resources to be moved more quickly around the world to keep cities functioning. Waste disposal systems were also designed to remove hazardous materials from the urban environment. This helped to improve public health within cities, but outputs of refuse, sewage and air pollution were usually displaced to another part of the environment. Sewers released human and industrial wastes into rivers, causing pollution downstream. Industrial smokestacks spewed air pollutants high into the atmosphere, dispersing harmful substances downwind. Refuse was incinerated or dumped 'out of sight and out of mind' in sanitary landfills beyond the boundaries of the city. Such approaches to waste disposal have proven to be inadequate and short-sighted, and attention has now shifted to preventative measures such as minimising wastes at source and maximising recycling. For long-term sustainability, cities must function with a 'circular metabolism' in which wastes are reused rather than the current 'linear metabolism' that simply displaces wastes elsewhere.

At the start of the twenty-first century, however, many cities in the developing world were still struggling to provide their poorest inhabitants with access to safe water supplies and proper sanitation. One of the key lessons of environmental history is that social inequality and ecological degradation often go hand in hand. Another is the interconnectedness of environmental crises, with a host of local and regional-scale problems cumulatively threatening to overwhelm the ecosystems that support life and maintain a habitable planet. Over the past five centuries (and particularly after 1950), escalating human activities – farming, industry and urbanism above all – have seriously degraded forest, grassland, riverine, coastal and other ecosystems. Anthropogenic climate change exemplifies how environmental issues can connect to become genuinely global in scale. According to the World Resources Institute, in 2000 agriculture was responsible for almost 14 per cent of greenhouse gas emissions; transport (especially private automobile use) also emitted nearly 14 per cent; industry and industrial processes produced just under 18 per cent; changing land use (primarily deforestation) just over 18 per cent; while energy consumption for residential and commercial buildings generated more than 33 per cent of the total. Because most humans now live and work in cities, and the world's urban population will continue to grow in the future, they are the major players in the climate change and sustainability arenas.

At the global level, between 1850 and 2000 the USA, Britain, Germany, Japan and other industrialised countries accounted for about 75 per cent of greenhouse gas emissions. But industrialising countries such as China and India – with their rapidly growing populations and booming economies – will drive carbon-dioxide build-up in the atmosphere for decades to come. Effective international policies and substantial investment in clean, renewable technologies are urgently needed to tackle climate change. However, the globalisation of economic life, with China and other developing countries manufacturing low-priced goods for export to the West (developed nations have essentially 'outsourced' their pollution), and the great expense of designing and disseminating environmentally sound technologies, makes negotiations to radically reduce emissions of greenhouse gases complex. Huge disparities in average carbon-dioxide emissions per person between developed and developing countries – over 19 tonnes in the USA, around 5 tonnes in China, and less than 2 tonnes in India – also means that a fair and equitable agreement to combat climate change will be difficult to reach. The matter is further complicated by the need to regulate the activities of large multinational corporations (some more powerful than small nation-states). But the cost of failure, as the former World Bank economist Nicholas Stern has warned, could be catastrophic. As well as the human suffering caused by the increased incidence of severe floods, droughts and food shortages, if global warming is left unchecked it could shrink the world's economy by up to 20 per cent. The impacts of climate change, though, will not be evenly distributed. 'The poorest countries and people', Stern says, 'will suffer earliest and most.'

To date, however, progress towards a binding global pact to cut greenhouse gas emissions has been slow. States and societies still prioritise economic growth over protecting the environment. A mechanism for discussion, the Framework Convention on Climate Change, was put in place following the 1992 United Nations Conference on Environment and Development – better known as the Earth Summit – held in Rio de Janeiro. But the resulting Kyoto Protocol, signed in Japan in 1997, which committed participating nations to a collective 5.2 per cent reduction in carbon-dioxide emissions (against 1990 levels) by 2012, was flawed and ineffective, with no tough penalties for non-compliance. Nonetheless, the USA withdrew from the process in 2002, worried about damage to its economy. More than half of the countries that signed are unlikely to meet their modest reduction targets. Although both China and India were involved, they did not have to curtail their emissions. If current 'Beyond Kyoto' negotiations do not produce a more robust international agreement to considerably reduce greenhouse gases, which is strictly enforced, global warming may become irreversible. Indeed, the increasingly heavy demands being placed on degraded ecosystems are diminishing the prospects for long-term sustainability, which many green critics have maintained is incompatible with economic development. Even if growth was halted now, and consumption levels scaled down in the affluent West, population expansion in the developing world (the planet is predicted to be home to some 9 billion people by 2050) will increase the pressure on natural systems.

On a more optimistic note, it is not too late to act. Earlier international environmental initiatives have enjoyed a modicum of success, such as the IUCN Convention on International Trade in Endangered Species of Wild Fauna and Flora (1975), the International Whaling Commission's moratorium on commercial whaling (1986), the United Nations Environment Programme's Montreal Protocol on Substances that Deplete the Ozone Layer (1987) and the designation of biosphere reserves under the UNESCO Man and the Biosphere Programme (there are now 531 sites worldwide in 105 countries). Turning to population growth, it is likely that modern farming methods and biotechnology (the genetic enhancement of crops) will avert a 'Malthusian crunch' for decades to come. However, while technically and politically possible, marrying the alleviation of hunger and poverty with the long-term goal of sustainable development will be no easy task. It will require a strong commitment to international cooperation; perhaps even political evolution towards global governance. A sustainable future will also mean unpopular lifestyle changes, such as curbing our 'throwaway' culture, reducing energy use, eating less meat and making fewer car and plane journeys. Unless environmental considerations begin to take precedence over material growth, we risk pushing the earth's ecosystems beyond their limits.

Further reading

Paul J. Crutzen, 'Geology of Mankind: The Anthropocene', *Nature* (vol. 415, January 2002, p. 23).

Rashid Hassan, Robert Scholes and Neville Ash (eds), *Millennium Ecosystem Assessment, Volume 1: Ecosystems and Human Well-Being – Current State and Trends* (Washington, DC: Island Press, 2005).

J. Donald Hughes, 'Global Environmental History: The Long View', *Globalizations* (vol. 2, December 2005, pp. 293–308).

John McNeill, *Something New under the Sun: An Environmental History of the Twentieth Century* (London: Allen Lane, 2000).

Clive Ponting, *A New Green History of the World: The Environment and the Collapse of Great Civilisations* (London: Vintage Books, 2007).

Nicholas Stern, *The Economics of Climate Change: The Stern Review* (Cambridge: Cambridge University Press, 2007).

UN-HABITAT, *State of the World's Cities 2008/09: Harmonious Cities* (London: Earthscan, 2008).

Worldwatch Institute, *State of the World 2009: Into a Warming World* (New York: W. W. Norton, 2009).

Index